# Specification for Tunnelling

# Specification for Tunnelling
## Fourth edition

**The British Tunnelling Society**

**Published by Emerald Publishing Limited**, Floor 5, Northspring, 21-23 Wellington Street, Leeds LS1 4DL.

ICE Publishing is an imprint of Emerald Publishing Limited

**Other ICE Publishing titles:**
*Monitoring Underground Construction: A best practice guide*
The British Tunnelling Society. ISBN 978-0-7277-4118-9
*Rock Engineering*
Håkan Stille and Arild Palmström. ISBN 978-0-7277-4083-0
*Tunnel Lining Design Guide*
The British Tunnelling Society and the Institution of Civil Engineers.
ISBN 978-0-7277-2986-6

A catalogue record for this book is available from the British Library

ISBN 978-0-7277-6643-4

Cover photo: Crossrail tunnel. Paul Daniels/Alamy Stock Photo

Commissioning Editor: Viktoria Hartl-Vida
Assistant Editor: Cathy Sellars
Production Editor: Sirli Manitski

Typeset by Manila Typesetting Company
Index created by LNS Indexing

# Foreword

The *Model Specification for Tunnelling* was produced by the British Tunnelling Society in conjunction with the Ground Board of the Institution of Civil Engineers in order to establish a common standard for tunnelling and first published in 1997. It was revised in 2000, published as the *BTS Specification for Tunnelling* and has become the standard Specification referred to in the majority of tunnelling contracts in the UK and in many worldwide. The third edition was published in 2010 and formed the basis of specifications for major infrastructure projects such as Crossrail, Thames Tideway and HS2.

This fourth edition consolidates the experiences gained on these landmark projects and presents the updated guidance emerging from practical application of the third edition. Like the earlier editions it relies heavily on the experience of individual and corporate members of the BTS. There have been too many contributors to name individually here.

Similar to the third edition, there are major changes in the section on sprayed concrete which reflect its growth in use and development as a technique. The sections dealing with then new techniques, such as sprayed applied waterproofing membranes, have been critically reviewed and likewise reflect practical experience. All sections have been updated to reflect best current practice, changes in national standards and the new Eurocodes. We trust that the fourth edition of the BTS Specification will retain its position as trusted standard for tunnelling contracts in coming years.

We recognise that change will continue to occur in the tunnelling industry and that it will always be possible to improve this specification. Any suggestions for improvements and amendments to future editions should be sent to the Secretary of the British Tunnelling Society at the Institution of Civil Engineers, Great George Street, London SW1P 3AA.

**Fourth Edition Drafting Committee**

Chair: Christoph Eberle, Mott MacDonald

Charles Allen, OTB Concrete

Divik Bandopadhyaya, London Bridge Associates

Mehdi Hosseini, Balfour Beatty

Benoit Jones, Consultant

Mike King, Consultant

Donald Lamont, Consultant

Omar Mohammed, J. Murphy & Sons

Richard Sutherden, J. Murphy & Sons

# Contents

**The British Tunnelling Society**
ISBN 978-0-7277-6643-4
https://doi.org/10.1680/st.66434.001

# 1. General requirements

# 101. General notes

## 101.1. General

1. This Specification is a model document intended to serve as a basis for materials and workmanship quality requirements for all tunnel projects, including adits and shafts, in conjunction with the specification component of the Institution of Civil Engineers Specification for Piling and Embedded Retaining Walls, and the scope can be extended to include cut-and-cover tunnels and similar underground structures.

2. This Specification is written in modular form. It is intended that the whole document be incorporated into documents by reference, and additional, substituted or deleted clauses particular to the Project be included in a Particular Specification. References are made to the Particular Specification at various points in the text for the user to insert project-specific requirements. Where the Particular Specification is in conflict with the general text of the Specification, the Particular Specification shall take precedence.

3. The Specification indicates minimum standards of materials and workmanship, and is written so that the parties to the Contract are as free as possible to agree on methods of carrying out the work.

4. Any clauses relating to work or materials not required in the project are deemed not to apply.

5. Whenever possible, reference has been made to other industry standards and accepted quasi-standards. Except in this chapter and where specific to the text, references to Acts of Parliament and Statutory Instruments have not been made, as compliance with legislation is a statutory requirement.

6. This Specification reflects tunnelling practice as undertaken by UK clients, contractors and designers, but has been written to allow it to be used also in an international context with the minimum of modification.

7. This Specification has been written so that it may be used with a range of procurement methods and Conditions of Contract. The title 'Engineer' has been used throughout this Specification as the accountable party who is empowered to make decisions on design and technical matters and variations, but this title will vary between procurement methods and Conditions of Contract. The Contract Documents or Particular Specification should set out the roles and responsibilities for the particular project.

8. The title 'Designer' has been used where it is essential that supervision or specific input is provided that fully understands the design basis. The 'Designer' may be employed by the Employer or the Contractor depending on the particular project and procurement method.

9. Assessment of payment and sharing of financial risk have been specifically excluded from this Specification, this being deemed to be the province of other Contract Documents.

10. This Specification is written to be used in conjunction with other Specifications where several disciplines are involved in the works. The Particular Specification should establish the order of precedence where more than one Specification is referenced.

11. This Specification has been written on the basis that details of the Contractor's methods and temporary works will be submitted to the Engineer for agreement. The Particular Specification should clarify the requirements for the Engineer's agreement to methods and temporary works for the particular project, including timescales for review. Where the Engineer's agreement to methods and temporary works is not required, this shall be stated in the Particular Specification.

12. The planning and implementation of the works shall comply with the current version of the International Tunnel Insurance Group *A Code of Practice for Risk Management of Tunnel Works* 2006 (being revised).

13. The Contractor shall be the *user* in terms of BS EN 16191 and shall be responsible for providing all information on *intended use* required by that standard.

14. In this specification 'Employer' means the entity commissioning and paying for the tunnel works.

**101.2. Carbon management**

1. The Designer and the Contractor shall manage and minimise whole life carbon across the delivery of the Contract, in line with the principles of the BSI Publicly Available Specification (PAS) 2080 *Carbon Management in Infrastructure*.

*Note: Due to the rapid development cycle of carbon management it is recommended to review the requirements of Clause 101.2 and associated clauses and confirm their applicability in the Particular Specification.*

2. The PAS 2080 carbon management hierarchy shall be applied for the management process as per Clause 101.2.1 to identify whole life carbon solutions for the asset and its components. In applying the carbon reduction hierarchy, all value chain members shall demonstrate they have considered the following

   (*a*) avoid: ensure the outcomes of the project/programme of work are aligned with the net zero transition and evaluate the basic need of an asset/network

   (*b*) switch: assess alternative solutions/options to reduce whole life emissions through alternative materials, technologies for operational carbon reduction, among

others, while satisfying the whole life performance
requirements

(c) improve: identify solutions and techniques that improve
the use of resources and design life of an asset/network,
including applying circular economy principles to assess
materials/products in terms of reuse and recycling
potential after end of life as well as flexibility in being
repurposed.

3. The Designer and Contractor shall identify carbon reduction
opportunities at the earliest stage of the project or programme
of work to enable the greatest carbon savings to be achieved.
The baseline against which savings are measured shall be
agreed in the Contract.

4. The definition of the goal and scope of greenhouse gas (GHG)
quantification shall capture the elements listed in PAS 2080
Section 7.1.1 and shall be agreed with the Engineer.

5. Carbon management shall be captured in the Quality
Management System as per Clause 108 of this specification.

6. For the use of non-standardised materials please refer to
Clause 103.3 of this specification.

## 102. Definitions

**102.1. General**

1. Where definitions are not provided within the specification, they shall generally be those contained in BS 6100.

2. Definitions of tunnelling terms are contained in BS 6100-3: Section 6 (2007).

3. For terms not defined in BS 6100, reference shall be made to the ITA 'Glossary of tunnelling terminology' (https://tunnel.ita-aites.org/en/component/seoglossary/1-main-glossary).

4. 'Mechanised tunnelling' refers to tunnels excavated by some form of tunnel boring machine.

5. 'Conventional tunnelling' refers to tunnels excavated by means other than tunnel boring machine – for example, drill and blast in rock.

6. 'Permanent works' are works, the purpose of which extends beyond completion of construction and/or are left in place beyond the completion of construction.

7. 'Temporary works' are works, the purpose of which does not extend beyond completion of construction.

8. 'Carbon' is the short-hand colloquial term used for GHGs, typically the six covered by the Kyoto Protocol – carbon dioxide ($CO_2$), methane ($CH_4$), nitrous oxide ($N_2O$), hydrofluorocarbons (HFCs), perfluorocarbons (PFCs) and sulphur hexafluoride ($SF_6$).

## 103. References to Standards

### 103.1. Standards and alternative standards

Materials, equipment and methods shall comply with the Standards and Codes of Practice indicated in the text. Where references in this specification are dated, the dated version of that document shall be used. Where references are undated, then the versions current at the date for submission of tenders shall be used.

For BS EN 16191 and BS EN 12110, the version used shall be that current when the declaration of conformity is made by the manufacturer.

The Contractor may propose the adoption of alternative standards and shall provide explanations with any proposals. The use of such standards shall be subject to the agreement of the Engineer.

### 103.2. Alternative materials and equipment

The Contractor may propose alternative materials or equipment to those specified provided either

(*a*) they are of at least equal quality and performance or
(*b*) they are of like quality and performance and comply with approved alternative standards.

If alternative materials or equipment are proposed, the Contractor shall submit comprehensive details including technical descriptions, drawings and specifications to demonstrate that the alternative complies with either requirement of this Clause. The adoption of such alternative materials or equipment shall be subject to the agreement of the Engineer.

### 103.3. Design with non-standardised materials

Where proposed alternative materials are not standardised, or are used within the context of a design standard not specifically developed for use with these alternative materials, the design shall be undertaken as 'Design assisted by testing' to BS EN 1990:2002 +A1:2005 Annex D.

## 104. Eurocodes and European and British Standards

**104.1. Eurocodes**

The following design standards (Eurocodes) are relevant and are referred to in the text

| | |
|---|---|
| BS EN 1990 | Basis of structural design (Eurocode 0) |
| BS EN 1991 | Actions on structures (Eurocode 1) |
| BS EN 1992 | Design of concrete structures (Eurocode 2) |
| BS EN 1993 | Design of steel structures (Eurocode 3) |
| BS EN 1995 | Design of timber structures (Eurocode 5) |
| BS EN 1997 | Geotechnical design (Eurocode 7) |
| BS EN 1998 | Design of structures for earthquake resistance (Eurocode 8) |

**104.2. British, European and International Standards**

The British designated version of the following European standards dual (BS EN) or dual triple numbered European and International standards (BS EN ISO) and their UK National Annexes are relevant and are referred to in the text

| | |
|---|---|
| BS EN ISO 62 | Plastics. Determination of water absorption |
| BS EN 196-1 | Methods of testing cement. Determination of strength |
| BS EN 196-2 | Methods of testing cement. Chemical analysis of cement |
| BS EN 197-1 | Cement. Composition, specifications and conformity criteria for common cements |
| BS EN 206 | Concrete. Specification, performance, production and conformity |
| BS EN 295-7 | Vitrified clay pipe systems for drains and sewers – Requirements for pipes and joints for pipe jacking |
| BS EN 335 | Durability of wood and wood-based products. Use classes: definitions, application to solid wood and wood-based products |
| BS EN 338 | Structural timber. Strength classes |
| BS EN 450-1 | Fly ash for concrete. Definition, specifications and conformity criteria |
| BS EN 471 | High-visibility clothing |
| BS EN 480 | Admixtures for concrete, mortar and grout. Test methods |
| BS EN ISO 527-3 | Plastics. Determination of tensile properties. Test conditions for films and sheets |
| BS EN 681-1 | Elastomeric seals. Material requirements for pipe joint seals used in water and drainage applications – Vulcanized rubber |
| BS EN 681-2 | Elastomeric seals. Material requirements for pipe joint seals used in water and drainage applications. Thermoplastic elastomers |

| BS EN 682 | Elastomeric seals. Materials requirements for seals used in pipes and fittings carrying gas and hydrocarbon fluids |
| BS EN 932 | Tests for general properties of aggregates |
| BS EN 933 | Tests for geometrical properties of aggregates |
| BS EN 934-2 | Admixtures for concrete, mortar and grout – Part 2: Concrete admixtures – Definitions, requirements, conformity, marking and labelling |
| BS EN 1008 | Mixing water for concrete – Specification for sampling, testing and assessing the suitability of water, including water recovered from processes in the concrete industry, as mixing water for concrete |
| BS EN 1011-1 | Welding. Recommendations for welding of metallic materials – General guidance for arc welding |
| BS EN 1011-2 | Welding. Recommendations for welding of metallic materials. Arc welding of ferritic steels |
| BS EN 1062-7 | Paints and varnishes. Coating materials and coating systems for exterior masonry and concrete. Determination of crack bridging properties |
| BS EN 1090-2 | Execution of steel structures and aluminium structures – Technical requirements for steel structures |
| BS EN 1097 | Tests for mechanical and physical properties of aggregates |
| BS EN 1367 | Tests for thermal and weathering properties of aggregates |
| BS EN ISO 1461 | Hot dip galvanized coatings on fabricated iron and steel articles. Specifications and test methods |
| BS EN 1537 | Execution of special geotechnical work – rock anchors |
| BS EN 1542 | Products and systems for the protection and repair of concrete structures. Test methods. Measurement of bond strength by pull-off |
| BS EN 1562 | Founding. Malleable cast irons |
| BS EN 1563 | Founding. Spheroidal graphite cast irons |
| BS EN 1744 | Tests for chemical properties of aggregates |
| BS EN 1849-2 | Flexible sheets for waterproofing. Determination of thickness and mass per unit area – Plastics and rubber sheets for roof waterproofing |
| BS EN 1916 | Concrete pipes and fittings, unreinforced, steel fibre and reinforced |

| | |
|---|---|
| BS EN 1928 | Flexible sheets for waterproofing. Bitumen, plastic and rubber sheets for roof waterproofing. Determination of watertightness |
| BS EN ISO 2286 | Rubber- or plastics-coated fabrics. Determination of roll characteristics |
| BS EN ISO 3506-2 | Mechanical properties of corrosion-resistant stainless steel fasteners. Nuts with specified grades and property classes |
| BS EN ISO 4624 | Paints and varnishes. Pull-off test for adhesion |
| BS EN ISO 8501-1 | Preparation of steel substrates before application of paints and related products. Visual assessment of surface cleanliness – Rust grades and preparation grades of uncoated steel substrates and of steel substrates after overall removal of previous coatings |
| BS EN ISO 9001 | Quality management systems. Requirements |
| BS EN ISO 9863-1 | Geosynthetics. Determination of thickness at specified pressures – Part 1: Single layers |
| BS EN ISO 9864 | Geosynthetics. Test method for the determination of mass per unit area of geotextiles and geotextile-related products |
| BS EN 10025 | Hot rolled products of structural steels. Technical delivery conditions |
| BS EN 10080 | Steel for the reinforcement of concrete. Weldable reinforcing steel. General |
| BS EN 10164 | Steel products with improved deformation properties perpendicular to the surface of the product – technical delivery conditions |
| BS EN 10226 | Pipe threads where pressure tight joints are made on the threads |
| BS EN ISO 11124 | Preparation of steel substrates before application of paints and related products (various dates) |
| BS EN ISO 11925-2 | Reaction to fire tests. Ignitability of products subjected to direct impingement of flame – Part 2: Single-flame source test |
| BS EN 12110 | Tunnelling machines – Air locks – Safety requirements |
| BS EN 12111 | Tunnelling machines – Roadheaders and continuous miners. Safety requirements |
| BS EN ISO 12236 | Geosynthetics – Static puncture test (CBR test) |
| BS EN 12310-2 | Flexible sheets for waterproofing. Determination of resistance to tearing (nail shank). Plastic and rubber sheets for roof waterproofing |
| BS EN 12317-2 | Flexible sheets for waterproofing. Plastic and rubber sheets for roof waterproofing |
| BS EN 12350 | Testing fresh concrete |

| | |
|---|---|
| BS EN 12390 | Testing hardened concrete |
| BS EN 12504-1 | Testing concrete in structures – Part 1: Cored specimens – Testing, examining and testing in compression |
| BS EN 12588 | Lead and lead alloys. Rolled lead sheet for building purposes |
| BS EN 12620 | Aggregates for concrete |
| BS EN 12878 | Pigments for the colouring of building materials based on cement and/or lime. Specifications and methods of test |
| BS EN 12889 | Trenchless construction and testing of drains and sewers |
| BS EN ISO 12958 | Geotextiles and geotextile-related products – Determination of water flow capacity in their plane |
| BS EN 13055 | Lightweight aggregates |
| BS EN 13139 | Aggregates for mortar |
| BS EN 13263 | Silica fume for concrete |
| BS EN 13491 | Geosynthetic barriers – Characteristics required for use in the construction of tunnels and associated underground structures |
| BS EN 13501-1 | Fire classification of construction products and building elements. Classification using data from reaction to fire tests |
| BS EN 13670 | Execution of concrete structures |
| BS EN 13791 | Assessment of in-situ compressive strength in structures and pre-cast concrete components |
| BS EN 14487-1 | Sprayed concrete – Part 1: Definitions, specifications and conformity |
| BS EN 14487-2 | Sprayed concrete – Part 2: Execution |
| BS EN 14488-1 | Testing sprayed concrete – Part 1: Sampling fresh and hardened concrete |
| BS EN 14488-2 | Testing sprayed concrete – Part 2: Compressive strength of young sprayed concrete |
| BS EN 14488-3 | Testing sprayed concrete – Part 3: Flexural strengths (first peak, ultimate and residual) of fibre reinforced beam specimens |
| BS EN 14488-4 | Testing sprayed concrete – Part 4: Bond strength of cores by direct tension |
| BS EN 14488-5 | Testing sprayed concrete – Part 5: Determination of energy absorption capacity of fibre reinforced slab specimens |
| BS EN 14488-7 | Testing sprayed concrete – Part 7: Fibre content of fibre reinforced concrete |
| BS EN 14651 | Test method for metallic fibre concrete. Measuring the flexural tensile strength (limit of proportionality (LOP), residual) |

| | |
|---|---|
| BS EN 14889-1 | Fibres for concrete – Part 1: Steel fibres. Definitions, specifications and conformity |
| BS EN 14889-2 | Fibres for concrete – Part 2: Polymer fibres. Definitions, specifications and conformity |
| BS EN 15167-1 | Ground granulated blast furnace slag for use in concrete, mortar and grout – definitions, specifications and conformity criteria |
| BS EN 16191 | Tunnelling machinery. Safety requirements |
| BS EN 16228 | Drilling and foundation equipment. Safety |
| BS ISO 18758 | Mining and earth-moving machinery. Rock drill rigs and rock reinforcement rigs – Part 1: Vocabulary |
| BS EN ISO 17855 | Plastics. Polyethylene (PE) moulding and extrusion materials |
| BS EN 60204 | Safety of machinery. Electrical equipment of machines |
| BS EN 61672-1 | Electroacoustics. Sound level meters. Specifications |
| ISO 9000 | Quality management systems |
| ISO 10406 | Fibre-reinforced polymer (FRP) reinforcement of concrete |
| PD CEN/TS 14416 | Geosynthetic barriers. Test method for determining the resistance to roots |

## 104.3. British Standards

The following British Standards are relevant and are referred to in the text

| | |
|---|---|
| BS 143 and 1256 | Threaded pipe fittings in malleable cast iron and cast copper alloy |
| BS 1134 | Assessment of surface texture. Guidance and general information |
| BS 1916 | Limits and fits for engineering |
| BS 4190 | ISO metric black hexagon bolts, screws and nuts. Specification |
| BS 4449 | Steel for the reinforcement of concrete – Weldable reinforcing steel – Bar, coil and decoiled product |
| BS 4482 | Steel wire for the reinforcement of concrete products. Specification |
| BS 4483 | Steel fabric for the reinforcement of concrete |
| BS 5228-1 | Code of practice for noise and vibration control on construction and open sites. Noise |
| BS 5228-2 | Code of practice for noise and vibration control on construction and open sites. Vibration |
| BS 5607 | Code of practice for the safe use of explosives in the construction industry |

| | |
|---|---|
| BS 5911-1 | Concrete pipes and ancillary concrete. Unreinforced and reinforced concrete pipes (including jacking pipes) and fittings with flexible joints (complementary to BS EN 1916:2002). Specification |
| BS 5975 | Code of practice for temporary works procedures and the permissible stress design of falsework |
| BS 6100 | Building and civil engineering. Vocabulary (various dates) |
| BS 6164 | Health and safety in tunnelling in the construction industry – Code of Practice |
| BS 6319 | Testing of resin and polymer/cement compositions for use in construction (various dates) |
| BS 6472 | Guide to evaluation of human exposure to vibration in buildings |
| BS ISO 4866 | Mechanical vibration and shock – Vibration of fixed structures – Guidelines for the measurement of vibrations and evaluation of their effects on structures *(Replaces BS 7385-1)* |
| BS 7371-8 | Coatings on metal fasteners. Specification for sherardised coatings |
| BS 7385-2 | Evaluation and measurement for vibration in buildings – Part 2: Guide to damage levels from groundborne vibration |
| BS 7668 | Weldable structural steels. Hot finished structural hollow sections in weather resistant steels. Specification |
| BS 7671 | Requirements for electrical installations. IET wiring regulations |
| BS 7973-1 | Spacers and chairs for steel reinforcement and their specification. Product performance requirements |
| BS 7973-2 | Spacers and chairs for steel reinforcement and their specification. Fixing and application of spacers and chairs and tying of reinforcement |
| BS 7979 | Specification for limestone fines for use with Portland cement |
| BS 8081 | Code of practice for rock anchors |
| BS 8102 | Code of practice for protection of below ground structures against water from the ground |
| BS 8500-1 | Concrete – Complementary British Standard to BS EN 206. Method of specifying and guidance for the specifier |
| BS 8500-2 | Concrete. Complementary British Standard to BS EN 206. Specification for constituent materials and concrete |
| BS 8666 | Scheduling, dimensioning, bending and cutting of steel reinforcement for concrete. Specification |

## 105. Other standards and documents

**105.1. Standards referred to in the text**

German Standards: DIN

| | |
|---|---|
| DIN 16726:2017 | Plastic roofing felt and waterproofing sheet; testing |
| DIN 53363:2003 | Testing of plastic films – Tear test using trapezoidal test specimen with incision |
| DIN 53861-1:1992 | Testing of textiles; vaulting test and bursting test; definitions of term |

American Standards: ASTM

| | |
|---|---|
| A820 | Standard Specification for Steel Fibers for Fiber-reinforced Concrete |
| A1011 | Standard Specification for Steel, Sheet and Strip, Hot-Rolled, Carbon, Structural, High-Strength Low-Alloy, High-Strength Low-Alloy with Improved Formability, and Ultra-High Strength |
| ACI 506R-03 | Guide to Shotcrete |
| ACI PRC-506-16 | Guide to Shotcrete |
| D5641-16 | Standard Practice for Geomembrane Seam Evaluation by Vacuum Chamber |
| D7361-07 | Standard Test Method for Accelerated Compressive Creep of Geosynthetic Materials |

Publicly Available Standards: PAS

| | |
|---|---|
| PAS 2080 | Carbon management in infrastructure |
| PAS 8820 | Construction materials. Alkali-activated cementitious material and concrete. Specification |

**105.2. Documents referred to in the text**

1. Building Research Establishment Digest 330 *Alkali–silica reaction in concrete (2004 edition) – Simplified guidance for new construction using normal reactivity aggregates*, 2004
2. Concrete Society Technical Report 31 *Permeability of site concrete*, 2008
3. fib Bulletin 83 Precast tunnel segments in fibre-reinforced concrete
4. Health and Safety Executive Guidance Note EH40 *Workplace exposure limits*, 2005
5. S1085 Fire Safety Performance of Materials
6. Pipe Jacking Association *Guide to best practice for the installation of pipe jacks and microtunnels*
7. Concrete Pipe Association *Concrete pipes for jacking small diameters (microtunnel) and unreinforced pipes*
8. The Fire Precautions (Sub-surface Railway Stations) (England) Regulations 2009

9. CIRIA PRJ PR 30 *Prediction and effects of ground movements caused by tunnelling in soft ground beneath urban areas 1996*

10. ISRM Document 2, Part 1 Suggested methods of rockbolt testing

11. International Society for Rock Mechanics – Document 2 Part 1 *Suggested methods of rockbolt testing*

12. German Concrete Association *Design principles of steel fibre reinforced concrete for tunnelling works 1992. Translation of DBV-Merkblätter Faserbeton – Technologie des Stahlfaserbetons und Stahlfaser Spritzbetons – Bemessungrundlagen für Stahlfaserbeton im Tunnelbau*

13. EFNARC Nozzleman Certification Scheme

14. Manual Handling Operations Regulations 1992

15. The Japan Society of Civil Engineers *SF4 Method of tests for flexural strength and flexural toughness of steel fiber reinforced concrete, Part 111-2 Method of tests for steel fiber reinforced concrete*, JSCE, 1984

16. CIRIA PRJ PR 30 Prediction and effects of ground movements caused by tunnelling in soft ground beneath urban areas 1996

17. Concrete Society Technical Report 63 *Guidance for the design of Steel-Fibre-Reinforced Concrete*

18. Burland JB, Broms BB, de Mello VFB Behaviour of foundations and structures, 1977

19. Allenby D and Ropkins, JWT IMechE lecture paper, 17 October 2007, London. *Jacked Box Tunnelling Using the Ropkins System™, a non-intrusive tunnelling technique for constructing new underbridges beneath existing traffic arteries*

20. National Structural Steelwork Specification for Building Construction (NSSSBC)

21. BRE Special Digest 1 *Concrete in aggressive ground*, 3rd edn, London, 2006

22. British Constructional Steelwork Association *National Structural Steelwork Specification for Building Construction*, 7th edn, London, 2020

23. Specification Series 1800 Structural Steelwork

24. British Tunnelling Society *The Management of Hand–Arm Vibration in Tunnelling. Guide to Good Practice*, London, 2006

25. Control of Substances Hazardous to Health Regulations (COSHH)

26. British Tunnelling Society *Occupational Exposure to Nitrogen Monoxide in a Tunnel Environment by the BTS Best Practice Guide*, London, 2008

27. STUVA (Studiengesellschaft für Tunnel und Verkehrsanlagen e.V.) Recommendation for gasket frames in segmental tunnel linings

28. Institution of Civil Engineers *Specification for Piling and Embedded Retaining Walls*, 3rd edn, ICE Publishing, London, 2016
29. Water Act 2003
30. Association of British Insurers & British Tunnelling Society *The Joint Code of Practice for Risk Management of Tunnel Works in the UK*, British Tunnelling Society, 2003
31. International Tunnel Insurance Group *A Code of Practice for Risk Management of Tunnel Works*, 2nd edn, 2012
32. Electricity at Work Regulations
33. Health and Safety Executive *Work in Compressed Air Regulations 1996* SI No. 1656, HMSO
34. Health and Safety Executive *A Guide to the Work in Compressed Air Regulations 1996*, HSE, Sudbury, 1996
35. Control of Pollution Act 1974
36. Health and Safety Executive *Safe use of vehicles on construction sites: A guide for clients, designers, contractors, managers and workers involved with construction transport* Guidance booklet 144, HSE, Sudbury, 2009
37. Environmental Protection Act 1990
38. Department for Transport Traffic safety measures and signs for road works and temporary situations – design. In *Traffic Signs Manual*, Ch. 8-1, 2nd edn, DfT, London, 2009
39. Control of Noise at Work Regulations 2005
40. ITA Working Group No. 5 *Guidelines for good working practice in high pressure compressed air*, ITA Report 10, ITA (International Tunnelling and Underground Space Association)/BTS (British Tunnelling Society)/CAWG (Compressed Air Working Group), 2nd revision, 2021
41. Control of Vibration at Work Regulations 2005
42. HSE EH75-2 *Occupational exposure limits for hyperbaric conditions – Hazard assessment document*, 2000
43. New Roads and Street Works Act 1991
44. Clayton CRI and Smith DM *Effective site investigation*, ICE Publishing, 2013
45. British Tunnelling Society *Monitoring Underground Construction – A best practice guide*, ICE Publishing, 2011
46. Mackenzie CNP *Traditional timbering in soft ground tunnelling – A historical review*, British Tunnelling Society, 2014
47. The Concrete Centre *National Structural Concrete Specification*, 4th edn, 2010
48. British Tunnelling Society Compressed Air Working Group *Guidance on good practice for Work in Compressed Air*, 2021
49. ITAtech Report 1, *ITAtech Guidelines on standard indication of load cases for calculation of rating life (l10) of TBM main bearings*, 2013

**50.** ITAtech Report 5, *ITAtech guidelines on rebuilds of machinery for mechanised tunnel excavation,* 2019

**51.** ITAtech Report 9, *ITAtech guideline for good practice of fibre reinforced precast segment – Vol. 2: Production aspects,* 2021

**52.** ITAtech Report 12, *ITAtech guidelines on services of machinery for mechanized tunnel excavation,* 2021

**53.** Österreichischer Betonverein (OEBV), *Guideline 'Tunnel Waterproofing',* 2015

**54.** Temporary Works Forum, *The use of European Standards for Temporary Works design,* see www.twforum.org.uk

**55.** Network Rail, NR/L2/CIV/044 Issue 4 *Planning, Design and Construction of Undertrack Crossings*

**56.** British Tunnelling Society *Guidance on the Work in Compressed Air Regulations* 1996

## 106. General provisions

The general provisions of the Contract shall be as stated in the Particular Specification.

This would normally include at least the following items

1. health, safety, welfare and environmental requirements
2. site areas
3. survey
4. fencing
5. levels and reference points
6. site accommodation
7. public relations
8. property interference
9. protection against damage
10. services
11. traffic
12. emergency arrangements
13. publicity
14. access to site
15. environmental impact
16. site working hours
17. security arrangements.

## 107. Occupational health, safety and welfare

**107.1. General principles**  The Contractor shall undertake the works in a manner that ensures so far as is reasonably practicable the health, safety and welfare of any person affected by the works. The measures to be taken shall include but not be limited to the adoption of systems of work, the use of plant, equipment and materials, the provision of adequate workplaces along with the means of access to and egress from them and a working environment that minimise the risks to the health, safety and welfare of all persons exposed.

**107.2. Statutory requirements**  All parties shall comply with their respective duties under the Health and Safety at Work etc. Act 1974, the Construction (Design and Management) Regulations 2015, the Work in Compressed Air Regulations 1996 and the other relevant statutory provisions as appropriate.

**107.3. Fundamental safety standards**  The Contractor shall comply in all respects with the following standards and guidance documents that are fundamental to health and safety in tunnelling. The Contractor shall also ensure compliance by other parties to the Contract, including Designers and Subcontractors as well as machinery, plant, equipment and materials suppliers.

| | |
|---|---|
| BS 6164 | Health and safety in tunnelling in the construction industry – Code of practice |
| BS EN 16191 | Tunnelling machinery – Safety |
| BS EN 12110 | Tunnelling machines – Air locks – Safety requirements |
| BS EN 12111 | Tunnelling machines – roadheaders, continuous miners and impact hammers – safety requirements |
| BTS | *Guidance on the Work in Compressed Air Regulations* 1996 |
| ITA | ITA Report 10 *Guidelines for good working practice in high pressure compressed air*, ITA (International Tunnelling and Underground Space Association)/BTS (British Tunnelling Society)/ CAWG (Compressed Air Working Group) |

**107.4. Employer's safety requirements**  The Contractor shall as a minimum

1. comply with the requirements of any relevant guidance, codes of practice or similar issued by the Employer for safe working
2. provide equipment for the rescue and evacuation of persons underground with persons instructed in its use to render assistance as required by the nature and scope of the Works
3. provide and maintain installation(s) for refuge for persons underground who are unable to escape to the surface in an emergency. The installation(s) shall be as required by the nature and scope of the Works, with all persons working

underground instructed and shown to be competent in their use. The installations shall conform with EN 16191 or ITA Report 14 unless otherwise instructed in the Contract

4. ensure all plant and equipment taken underground meets the requirements of BS 6164:2019. These requirements form a package of measures designed to mitigate the additional risks arising from work in the underground environment compared with surface use. These measures include but are not limited to the requirements set out in the sub-clauses 5–10 below

5. ensure all mobile and static plant over 5 kW in power shall be fitted with fixed fire suppression systems for class A and B fires. The suppression media shall provide kill and quench capability to inhibit re-ignition and cover at least the cab, engine/motor compartment, fuel tanks, hydraulic system and tyres. Class B fire extinguishing medium shall be used for suppression over all propulsion battery installations. Batteries based on lithium metal as opposed to lithium-ion technology shall not be used underground

6. ensure all mobile or static plant and equipment used underground shall use HFDU hydraulic fluid. This shall include use as transmission fluid

7. ensure all mobile plant and equipment, for which the operator position for travel and operation is on the machine, shall be fitted with a ventilated cab complying with BS EN ISO 23875:2022. The cab shall provide for driver protection and containment in the event of overturning as well as protection against falling objects, impact, dust, noise, vibration and heat. It shall not be possible to travel or operate the machine if the cab door is not properly shut

8. ensure all plant and equipment that is routinely mobile in use – for example, dumpers – shall be fitted with a lighting system that shows white in the direction of travel and red to the rear. Changeover shall be automatic on selection of the direction of travel

9. rubber tyred or tracked plant and machinery shall only be allowed underground if it has been designed, constructed or modified for such use. All plant and machinery taken underground shall fully comply with the requirements of BS 6164:2019 including

   (i). the use of HFDU fluids in hydraulic systems
   (ii). the use of fire-resistant hydraulic hoses
   (iii). the use of low smoke and fume cabling
   (iv). they have the protection of a comprehensive fire suppression system

> *(v).* vehicles that travel as part of their normal function shall have white lights to the front/red lights to the rear when travelling
>
> *(vi).* they shall have cabs that provide operator containment, do not impede operator visibility and protect the operator from dust, noise, impact, vibration, atmospheric contamination

**10.** mobile elevating work platforms used underground shall have overhoist protection when working under a supported roof and ISO 3449 FOPS level 1 protection otherwise. The machine shall be designed to withstand relevant FOPS impact loads.

**107.5. Emergency drills**

All safety and emergency procedure training shall be reinforced by regular practice drills.

**107.6. Shift length**

The Contractor shall take due account of local statutory restrictions, the Employer's requirements and the effects of fatigue on personnel in determining shift length.

**107.7. Occupational health**

1. The Contractor shall establish occupational health facilities on the surface, staffed by appropriate occupational health professionals on a full-time or part-time basis as required by the nature and scale of the Works.
2. Persons, when first employed on the site, shall be subject to appropriate pre-employment occupational health checks with periodic reassessment at intervals not exceeding two years.
3. The Contractor shall establish on site welfare and first aid facilities with appropriately trained personnel, both on the surface and underground, as required by the scale of the Works. Welfare facilities shall include changing and clothes drying/storage, toilet, and washing, showering and messing facilities. Where water washing facilities cannot be provided, appropriate alternative means of hand cleaning shall be provided. Barrier creams etc. for skin protection shall also be provided. There shall be a supply of cold potable water at the surface facilities and at clearly marked locations underground.

## 108. Quality Management and records

### 108.1. Quality Management System

1. The project shall be administered using an accredited Quality Management System conforming to BS EN ISO 9001. The individual requirements for agreement by the Engineer of materials and workmanship throughout this Specification shall be incorporated into agreed self-certification procedures.
2. The agreed quality control arrangements, including hold points and submission of records for the Engineer's acceptance, shall be set out in agreed Inspection and Test Plans.
3. A summary table of records for submission to the Engineer shall be included in the documentation of the quality control arrangements.

### 108.2. Engineer's agreement

References to the agreement of materials, workmanship, methods and so on throughout this Specification shall be interpreted as requiring the agreement of the Engineer.

### 108.3. Site records

1. The Contractor shall maintain all records necessary under this Specification, including quality records as appropriate.
2. Electronic records shall be maintained and backed up on a daily basis to prevent loss of data in the event of failure of electronic data storage.
3. The Contractor shall provide a Common Data Environment for storing site records and exchange of information on the project and be responsible for its maintenance until the end of the Contract. The structure for the Common Data Environment shall be agreed with the Engineer before the start of the project. All data exchange shall be undertaken to this structure unless agreed otherwise.
4. Site records shall be provided to the Engineer daily and copies made available to the Engineer at the end of the Contract, unless agreed otherwise.
5. All recorded information shall be documented in a humanly readable format in PDF and delivered to the Engineer electronically, unless agreed otherwise.
6. The ultimate owner of site records shall be specified in the Contract.
7. All site records shall be revision controlled. Communication between the Contractor and the Engineer shall reference the revision of any quoted documents.

### 108.4. As-built records

The Contractor shall supply the Engineer with all information necessary for the Health and Safety File including as-built drawings and records, maintenance schedules, operation and maintenance manuals, within the time specified in the Contract, after substantial completion of the Works. Information shall be provided in the agreed format. The Health and Safety File shall be prepared by the party identified in the Contract.

**108.5. Project BIM model**

1. The Contractor shall supply the Engineer with all the information necessary to produce a BIM model of the Works. The elements to be included in the BIM model and the level of detail of the BIM model shall be specified in the Contract.
2. The Contract shall specify the party responsible for production of the BIM model and the relevant standard requirements.
3. Specifically, the BIM model of the works shall include any concealed elements that remain in situ, the excavated profile and any ground treatment associated with the construction of the elements of the Works.

**108.6. Ongoing asset management**

The Works shall be inspected upon substantial completion, and findings retained and handed over to the Client as baseline for the ongoing asset management.

## 109. Competence

### 109.1. General

1. The Contractor shall be responsible for ensuring the competence of all staff under their control, including Subcontractors, working on the project.

   *Note: Employers have statutory duties in respect of the provision of information, instruction, training and supervision. Nothing in this section over-rides these statutory duties.*

2. Reference shall be made to clause 6.6 of BS 6164:2019 for a general statement on competence in tunnelling works.

### 109.2. Job-related competence assessment scheme

1. The Contractor shall set up and operate a scheme whereby all those working on the tunnelling project either for the Contractor or for Subcontractors demonstrate they are competent to undertake the job for which they are employed.
2. The scheme shall be based on an industry recognised competence framework such as the SKATE matrix (https://britishtunnelling.com/tts/skate-competence-matrix-for-tunnelling).
3. As a basic requirement for ensuring competence, the Contractor and Subcontractors shall only employ persons meeting the requirements for skills, knowledge and training set out in the SKATE matrix or industry accepted equivalent.
4. The Contractor or Subcontractor concerned shall also be able to demonstrate that the jobholder meets relevant experience criteria such as set out in the SKATE matrix.
5. Not all those working on the tunnelling project will necessarily fall into job categories set out in a tunnelling-specific competence matrix such as SKATE – for example, catering staff, administrative staff, occupational health staff, occupational hygienists, specialist technical advisers. The Contractor and Subcontractors shall put in place a system for such jobholders to demonstrate competence based on relevant vocational qualifications, professional memberships and experience.
6. The Contractor or Subcontractor concerned shall maintain records to demonstrate that a jobholder has demonstrated relevant competence before being put to work on the project as well as records of subsequent training and competence assessment.

### 109.3. Health and safety competence

1. Everyone working on site shall be able to demonstrate personal health and safety competence for surface construction works. Minimum requirements shall be to hold a CSCS card or C&G 6151 qualification.
2. Everyone working underground shall be able to demonstrate personal health and safety competence for working in the underground tunnelling environment.

**109.4. Induction training**

1. The Contractor shall set up and deliver an induction training programme for all those working on the tunnelling project. There may be a general site module for all jobholders, with a specific module only for those working underground.

2. The induction training programme shall as a minimum address the form of construction being undertaken on that site along with the hazards inherent in that form of construction and precautions to be taken, the welfare provisions available as well as emergency procedures including fire safety and first aid. Such induction shall be repeated whenever there is a material change in the working arrangements. The Contractor shall issue a written summary of the information provided to all persons working on the site.

3. The Contractor shall maintain a record of all persons inducted and the steps taken to demonstrate their competence and understanding of the information provided.

4. Separate arrangements may be made for visitors.

**The British Tunnelling Society**
ISBN 978-0-7277-6643-4
https://doi.org/10.1680/st.66434.027

# 2. Materials

## 201. General requirements

### 201.1. General

All materials supplied to the Works shall conform to all of the following

(*a*) this Specification
(*b*) the appropriate British or European Standard
(*c*) where an industry certification scheme is available, material supplied shall be supplied in accordance with that scheme
(*d*) materials shall be supplied from a quality assured source, operating a Quality Assurance system in compliance with the relevant part of BS EN ISO 9001
(*e*) the Construction Products Regulations, including the requirement for all products to be CE or UKCA marked
(*f*) the requirements of COSHH Regulation 7 to avoid the use of materials hazardous to health in the workplace.

### 201.2. Samples

Where required in the Particular Specification or where stated on the Drawings, samples should be supplied and the subsequent material shall conform to the samples.

### 201.3. Material use

Materials used on site shall be used in accordance with the supplier's recommendations and instructions.

### 201.4. Material storage and handling

All materials should be handled and stored so to maintain their integrity and to avoid damage and degradation.

### 201.5. Inspection and testing

Details of the level of inspection and testing to be adopted in respect of supplied materials shall be agreed with the Engineer prior to commencement of work.

### 201.6. Risk reduction in material selection

The Designer and the Contractor shall take account of any hazardous properties, chemical and mechanical, when specifying or selecting materials and shall select those that minimise the health and safety risk to those who come into contact with them and minimise any environmental risk.

### 201.7. Carbon reduction in material selection

The Designer and the Contractor shall optimise the design to achieve the lowest practical whole-life embedded carbon and shall select those design solutions that minimise the carbon footprint over the design life of the Works.

## 202. Concrete

### 202.2. General

1. All concrete shall be produced in accordance with BS EN 206 and BS 8500 unless where otherwise provided for in the Contract.
2. Where concrete is to be placed in aggressive ground, appropriate ground investigation shall be undertaken to identify the nature of the chemical composition of groundwater and ground.

### 202.2. Constituent materials – cement

1. The Contractor shall submit cement and cementitious material suppliers' certificates in accordance with the relevant British Standard. Details of all cements and cementitious materials shall be supplied including any alternative sources that might be used. The Contractor shall show that the quantity and quality required can be attained and maintained throughout the construction period.
2. CEM I will comply with BS EN 197.
3. Where sulphate resistance is required, the selected cement will be appropriate to the required Design Chemical (DC) class.
4. Cementitious materials shall have a reactive alkali content not exceeding a value of 0.6% by mass and/or the total mass of reactive alkali in the mix shall be calculated and controlled to satisfy the requirements of BS 8500-2 and the British Research Establishment (BRE) Digest 330. Certification will be supplied by the producer to demonstrate compliance with BRE Digest 330.
5. Blended cements and alkali-activated cementitious materials with proportions outside the limits specified in BS 8500-2 must be demonstrated by testing to meet the performance requirements to the satisfaction of the Engineer and Designer. The 'design assisted by testing' approach described in BS EN 1990:2002 +A1:2005 Clause 5.2 shall be used. As a minimum, alkali-activated cementitious materials should conform with PAS 8820.
6. Cementitious materials shall be delivered in bulk or in sealed and marked bags and shall be protected from the weather by enclosed transfer systems or other approved coverings. Cements that have exceeded the manufacturer's designated shelf life will not be used, and appropriate measures shall be taken for its safe disposal or return to the manufacturer.
7. A sample of every type of cement used shall be kept for Vicat testing (BS EN 196) if batching on site. When batching off site clear supplier records shall be kept and made available to the Engineer.

## 202.3. Constituent materials – aggregates

1. Aggregates shall conform to BS 8500-2. The Contractor shall obtain the agreement of the Engineer for the proposed aggregate sources and blends, and shall demonstrate compliance with laboratory tests that shall be carried out at monthly intervals to confirm the continued suitability of the aggregates.

   (*a*) Normal and heavyweight aggregates shall conform to BS EN 12620.
   (*b*) Lightweight aggregates shall conform to BS EN 13055-1.
   (*c*) Coarse recycled concrete aggregate (RCA) shall conform to BS 8500-2.

2. Aggregate shall be free from earth, clay, loam and soft, clayey, shaley or decomposed stone, organic matter and other impurities and shall be hard and dense.
3. Aggregates shall not contain any other matter likely to affect the long-term durability of the concrete. Reference is to be made to BRE Digest 330 for guidance in reducing the risk of deleterious alkali–silica reaction to the absolute minimum.
4. Tests shall be carried out in accordance with British Standards, as appropriate, and the results shall comply with the limits given therein, or as otherwise specified. Testing will be carried out to BS EN 932, BS EN 933, BS EN 1097 and BS EN 1744, as appropriate, and the results shall comply with the limits given therein and in BS 8500 and BS EN 206. Testing records shall be submitted to the Engineer once per month to confirm the suitability of the aggregates.
5. Crushed sand may be added to natural sand in approved proportions in order to achieve the required grading. When tested, the resultant material will comply with BS EN 12620.
6. Sand for mortars and grouts shall comply with BS EN 13139.
7. Coarse aggregate shall be as defined in BS EN 12620.
8. Coarse aggregate shall be tested for drying shrinkage characteristics in accordance with BS EN 1367-4. The drying shrinkage shall not exceed 0.075%.
9. The acid-soluble sulphate ($SO_3$) level shall not exceed the values specified in BS EN 12620.
10. The maximum permitted level of equivalent acid-soluble chloride ions ($Cl^-$) for any single constituent or combination of the constituents of the concrete in the hardened mix shall not exceed the limits given in BS EN 206.
11. The total estimated sulphate content ($SO_3$) shall comply with the limits given in BS EN 206.
12. Hardness and abrasion characteristics of the aggregate will comply with BS EN 12620.

13. Water absorption shall not exceed the permitted value in BS EN 12620.
14. Where specific thermal characteristics of the mix are required, the aggregate will be appropriately selected and tested in accordance with BS EN 1367.
15. Each size of aggregate shall be stored separately in drained concrete-based bins or on stages to prevent intermixing and the inclusion of foreign materials.
16. Incinerator bottom ash aggregate (IBAA) shall not be used in foam concrete.

**202.4. Constituent materials – water**

1. Water to be used for mixing and curing concrete and mortar shall be fresh and free from sediment and dissolved or suspended matter that may be harmful and shall comply with the requirements of BS EN 206 and BS EN 1008.
2. Recycled water may be used, provided controls are in place to demonstrate compliance with BS EN 206 and BS EN 1008.

**202.5. Constituent materials – admixtures**

1. All admixtures shall comply with BS EN 206 and BS EN 934.
2. Unless specified in the Contract, the use of set-retarding and water-reducing admixtures shall be subject to the agreement of the Engineer.
3. The Contractor shall determine the hazardous properties of admixtures during their selection and shall ensure appropriate health and safety measures are in place for their storage and use.

**202.6. Constituent materials – fibres**

Fibres shall comply with Clause 203.3.

**202.7. Constituent materials – additions**

1. General suitability as a Type II addition as defined in BS EN 206 is established for the following

   (a) fly ash conforming to BS EN 450-1
   (b) silica fume conforming to BS EN 13263
   (c) ground granulated blast furnace slag (GGBS) conforming to BS EN 15167
   (d) metakaolin with an appropriate Agrément certificate.

2. General suitability as a Type I addition as defined in BS EN 206 is established for the following

   (a) filler aggregate conforming to BS EN 12620 or BS EN 13055-1
   (b) pigments conforming to BS EN 12878.

General suitability of limestone fines conforming to BS 7979 is established for use in combinations conforming to BS 8500-2 Annex A.

**202.8. Concrete mixes**

1. The grade and properties of the concrete used in each part of the work shall be as stated on the Drawings or in the Particular Specification and shall be in accordance with BS 8500 and BS EN 206.
2. The selection design and quality control of mixes shall be carried out by the Contractor or on their behalf by the manufacturer.
3. If the finish of the concrete is required to be of a controlled or superior standard, then trial panels will be manufactured 35 days in advance of the Works starting, and the finish achieved will be approved by the Engineer before commencing work. The panel will be retained during the course of the Works to use as a comparative measure for the Works.
4. If existing data on materials and properties of trial concrete mixes are not available, preliminary laboratory tests shall be carried out to establish the mixes to satisfy the Specification with the available materials.
5. Laboratory trial mixes shall be tested to determine compliance with BS 8500 and BS EN 206 for all the required properties of the mix.
6. Unless otherwise agreed with the Engineer, field trial mixes shall be prepared under full-scale site conditions at least 35 days before the commencement of concreting and tested in accordance with BS EN 12350 and BS EN 12390.
7. The field trial mixes shall be tested to determine compliance under statistical evaluation where required by BS EN 206. An acceptable value for the limits of the required properties shall be established during the trials, which shall thereafter be used to monitor the quality control of the mixes and set the standard of compliance.

**202.9. Ready-mixed concrete**

1. Use of ready-mixed concrete and its source shall be subject to the prior agreement of the Engineer, and the Contractor shall use only third-party accredited quality assured companies.
2. Water shall not be added to concrete in a truck mixer drum other than at the batching plant, unless approved by the Engineer in a controlled manner under the supervision of the producer's representative and recorded on the delivery note. The mix shall be continuously agitated during transportation.
3. The transportation and placing times of ready-mixed concrete shall be reviewed in relation to all the circumstances including travel distance and risk of traffic delays *en route*. Unless special

measures are taken, the concrete will be placed in the Works within 2 h after addition of the water to the cement. The time between consecutive loads finishing placing and starting placing shall not exceed 30 min.

4. The Contractor shall provide certificates to demonstrate compliance of each component of the mix with the relevant clauses of the specification. The delivery note for each batch shall state the designation of the concrete mix, the type of cement and minimum cement content, the maximum aggregate size, the workability class of the mix, the chemical exposure class of the concrete, the admixtures used, the time that the concrete was mixed and the weight of the constituents of each mix along with any other specified requirements.

5. Concrete temperature at the time of delivery, when measured in accordance with BS 8500, shall not exceed any value specified by the Engineer in the Contract, or 35°C.

**202.10. Concrete batching**

1. Production control of concrete will satisfy the requirements of BS EN 206 and BS 8500-2.

2. All constituents shall be weighed or metered in accordance with the limits prescribed in BS EN 206.

3. Admixtures shall only be introduced using purpose-made equipment accurately calibrated. Where such equipment is unavailable, and where agreed with the Engineer, alternative dosing methods to the manufacturer's recommendations may be adopted.

4. Water shall not be added to concrete after it has left the mixer unless controlled, recorded and agreed with the Engineer.

5. Materials shall not be heated unless agreed with the Engineer.

6. Where fibre reinforcement is added to the concrete mix, this shall only be introduced using purpose-made equipment.

**202.11. Quality control**

1. The Contractor shall plan all concrete Quality Management procedures in accordance with Clause 201.5.

2. The conformity control of strength parameters required will be demonstrated in accordance with BS EN 206. Specimens tested to demonstrate compliance will be cubes, cylinders or prisms appropriate to the testing standards and BS EN 206.

3. Test samples shall be made, cured, stored, transported and tested to BS EN 12350 and BS EN 12390. Spot samples will not be used to evaluate strength parameters.

4. Concrete cube test results will be acceptable if statistical analysis of the results meets the requirements of BS EN 206.

5. Concrete shall be tested for durability properties by means of absorption and capillary suction (sorptivity) tests where appropriate. An appropriate test method will be agreed by all parties before testing is undertaken.

6. Compaction factor, slump, Vebe, flow table or other workability tests shall be carried out as required during concreting of permanent works to control workability at the batching plant and at the site of the pour. The degree of workability shall be as specified or as determined during the trial mixes; permitted tolerances shall be in accordance with BS EN 206. Samples tested will be either spot samples or composite samples taken in accordance with BS EN 12350-1 and the appropriate tolerances for compliance will be applied in each case.

7. Clear supplier records shall be kept of all constituent materials. The Contractor shall make these available to the Engineer.

8. Where the conformity control of specific mechanical concrete parameters is not standardised the Contractor's Quality Management plan shall contain provisions to demonstrate conformance of these specific parameters with the performance requirements to the satisfaction of the Engineer and Designer. The 'design assisted by testing' approach described in BS EN 1990:2002+A1:2005 Clause 5.2 shall be used.

9. The Contractor shall follow all specific pre-production and production control requirements identified by the Designer.

## 203. Reinforcement

### 203.1. Steel bar reinforcement

1. Reinforcement for use in reinforced concrete shall comply where appropriate with BS 4449, 4482, 4483 and BS EN 10080.
2. Reinforcement shall be obtained from a Certificated Authority for Reinforcing Steels (CARES) Quality Assurance approved supplier and the Contractor shall provide copies of the supplier's certificates of test results relating to the steel reinforcement to be supplied.
3. Reinforcement shall be stored in a clean, dry environment under cover and racked as necessary for protection from aggressive elements.
4. Steel reinforcement shall be cut and bent in accordance with BS 8666.
5. Methods of tying reinforcement that avoid the repetitive manual use of pliers shall be preferred. Tying wire shall be 1.6 mm diameter soft annealed mild steel, and when fixed shall not project into the concrete cover. Welding of reinforcement shall be undertaken in accordance with BS EN ISO 17660.
6. Where the Contract so requires, the Contractor shall produce bending schedules, prepared in accordance with BS 8666.
7. Reinforcement cages shall be strengthened as necessary to allow them to be lifted.
8. Bar reinforcement placed in holes drilled into concrete shall be fixed using a proprietary fast-setting resin in accordance with the Drawings and the manufacturer's recommendations. Polyester and hybrid polyester resins are prohibited in wet or underground locations. The quality of such connections shall be demonstrated by representative testing of the in situ connections to the satisfaction of the Engineer.

### 203.2. Welded wire fabric

Welded wire fabric shall comprise hard-drawn wire in accordance with BS 4482 and BS 4483. It shall be firmly fixed in place using an agreed method. Overlap between adjacent sheets of welded wire fabric shall be a minimum of 2 squares.

### 203.3. Fibre reinforcement

1. Fibre reinforcement shall be obtained from a supplier meeting the requirements of BS EN ISO 9001. Fibres used for the purpose of reinforcement shall be CE certified to Assessment and Verification of Constancy of Performance (AVCP) System 1.
2. Fibres are generally accepted for use in concrete conforming to BS EN 206 and BS 8500 if the fibre conforms to BS EN 14889.
3. Fibre-reinforced concrete shall be trialled and tested to ensure it meets the Designer's requirements before inclusion in the Works. Fibre-reinforced concrete performance testing shall be aligned with the design approach as specified by the Engineer. Historical data of the same fibre and dosage may be accepted in place of trials subject to confirmation by the Engineer.

4. Steel fibres shall be deformed steel fibre in accordance with BS EN 14889-1 Group 1 (cold-drawn steel).
5. Macro-synthetic fibres shall be in accordance with BS EN 14889-2.
6. Micro fibres and other fibres for non-structural purposes shall be CE certified to Assessment and Verification of Constancy of Performance (AVCP) System 3.
7. Alternative fibres may be utilised following successful completion of proving trials that meet the Designer's performance requirements.
8. Fibres shall be stored, handled, batched and mixed in accordance with the manufacturer's recommendations. Generally, this will require them to be stored in dry packaging until ready for use and shall be free from corrosion, oil, grease, chlorides, and deleterious materials that may reduce the efficiency of mixing or spraying processes, or that may reduce bond between the fibres and the concrete.
9. Fibre material, type and dosage shall be selected such that the performance requirements specified on the Drawings or in the Particular Specification are achieved. This shall be demonstrated by laboratory trials undertaken and agreed with the Engineer prior to commencement of concrete production.
10. Fibre type and dosage shall be selected considering ease of use in the batching, mixing and concrete placement processes proposed as demonstrated by site trials.
11. Fibres shall be added and mixed in a manner to produce a homogeneous distribution within the concrete matrix. Testing of concrete shall demonstrate that the fibres are being uniformly distributed throughout the concrete mix.

## 203.4. Non-metallic reinforcement

1. Non-metallic reinforcement shall comply with the requirements of ISO 10406.
2. Non-metallic reinforcement shall be obtained from a supplier meeting the requirements of BS EN ISO 9001 or similar.
3. The manufacturer shall declare as a minimum

   (a) relevant product standard
   (b) fibre and matrix materials
   (c) nominal dimensions such as diameter, width, thickness
   (d) characteristic tensile strength, modulus of elasticity and the ultimate strain in the direction of the fibres
   (e) glass transition temperature
   (f) bond characteristics.

4. Non-metallic reinforcement shall be stored, handled and used in accordance with the manufacturer's recommendations. Generally, this will require storage in a clean, dry environment under cover and racked as necessary for protection from aggressive elements.

## 204. Precast concrete linings

### 204.1. General

1. Precast concrete segments for linings shall be supplied by the Contractor from an agreed manufacturer, or manufactured in a suitable factory, for erection in the Works in accordance with the Contract requirements.
2. Segments of the lining and all accessories such as gaskets, bolts, dowels and packers shall be as specified on the Drawings.
3. Lining segments may be cast in existing moulds and be a manufacturer's standard design or be a particular design to suit the requirements of the project and tunnelling plant.
4. Concrete shall be in accordance with Clause 202.
5. Reinforcement shall be in accordance with Clause 203. Cages may be formed by welding or tying.
6. Details of the selected accessories such as gaskets, bolts, dowels and packers shall be supplied to the Engineer for agreement.

### 204.2. Manufacture of segments

1. Manufacturing facilities for segmental lining systems will be required to show

    (a) a certified Quality Assurance and control programme to BS EN ISO 9001 accepted by the Engineer
    (b) compliance with British Standards and this Specification regarding materials, mixing and placing, curing and storing of concrete constituents, concrete segments and fixings.

2. The manufacturer's premises and methods shall be open to inspection by the Engineer for the purpose of checking the quality of manufacture. The Contractor shall ensure that all necessary assistance is provided to the Engineer on each visit. Such visits shall not be taken as a substitute for the segment manufacturer's own quality checking.
3. The segment manufacturer shall produce records demonstrating the compliance of the cast segments with all associated requirements. The records shall be updated and shared with the Engineer on a weekly basis, or as agreed otherwise.
4. For new precasting segment manufacturing lines in an existing facility or new segment production facilities, full-scale production trial requirements shall be detailed in the Particular Specification prior to any segments being manufactured for use in the Works.

## 204.3. Moulds

1. Moulds shall be robustly constructed, tightly jointed and properly maintained such that the dimensions of the segments are always within the specified tolerances.
2. Where new moulds are being manufactured for the particular project, the fabrication Drawings shall be submitted to the Engineer for agreement.
3. Fabrication Drawings and details of the moulds to be used for casting concrete segments shall be supplied to the Engineer for their agreement before prototype segments are cast. Trial segments shall be made for the Engineer's inspection unless agreed otherwise. Samples shall be marked indelibly and set aside for reference purposes.
4. The manufacturer of the moulds shall select their manufacturing tolerances so that the production tolerances as per Clause 204.4 can be achieved.
5. Acceptance of the moulds shall be on the basis of compliance of the cast segment with the specified geometry, including tolerances.

## 204.4. Tolerances for the manufacture of bolted segments

### 204.4.1. Segments

Dimensions of individual precast concrete special segments shall be within the following tolerances. The tolerances apply to the finished concrete segments unless noted otherwise.

| | | |
|---|---|---|
| (a) | circumferential length | ±1 mm |
| (b) | thickness | −3 mm, +5 mm |
| (c) | width (length in the direction of the tunnel axis) | ±1 mm |
| (d) | square | diagonal dimension ±1 mm from theoretical dimension |
| (e) | bolt hole, guiding rod and dowel recess: diameter | +1 mm, −0.2 mm (or as specified by accessory manufacturer) |
| (f) | bolt hole, guiding rod and dowel position | ±1 mm |
| (g) | glued-in gasket groove: depth measured relative to joint contact face | ±0.25 mm |

| | | |
|---|---|---|
| (h) | glued-in gasket groove: width | ±0.25 mm |
| (i) | glued-in gasket groove: position | ±1.5 mm from specified position |
| (j) | longitudinal joints in a plane generally along the axis of the tunnel (longitudinal) | 0.3 mm from theoretical plane with rate of deviation not exceeding 1 mm/m |
| | in a radial plane *(Not applicable for convex joints)* | 0.2 mm from theoretical plane with rate of deviation not exceeding 1 mm/m |
| (k) | circumferential faces | 0.5 mm from theoretical plane with rate of deviation not exceeding 1 mm/m |
| (l) | smoothness of other faces | |
| | back face | smooth float ±1.5 mm |
| | front face | formed face ±1 mm |
| (m) | mismatch of sealing groove at corners | <0.5 mm |
| (n) | dowel axis divergence | maximum of 1 deg from the normal on the corresponding segment surface and 1mm. Value to be verified on moulds. |
| (o) | cast-in gasket: protrusion from joint contact surface | allowable protrusion from the joint contact surface shall be the sum total of the gasket fabrication tolerance, as accepted by the Designer, and a groove fabrication tolerance of +/– 0.25 mm |
| (p) | cast-in gasket: position | ±1.5 mm from specified position |
| (q) | location of erector cones, grouting ports, laser spots and other markers | ±2 mm |

**204.4.2. Rings**

At least two test rings shall be erected as a vertical stack on a flat and level base, in a form and sequence representative of the construction arrangement to be agreed with the Engineer. The frequency of test rings, if not specified within the Contract, is to be agreed with the Engineer prior to commencement of manufacture and can be increased if required during the manufacturing period. All gaskets, dowels and packings are to be removed from both rings. Alternative methods may be used where agreed with the Engineer including virtual ring build check. The following dimensions shall be checked

| | |
|---|---|
| (*a*) internal diameter (adjusted for packer removal where necessary) | rings up to 6 m internal diameter: ±6 mm rings exceeding 6 m internal diameter: ±10 mm |
| (*b*) lip between adjacent segments on internal diameter | <3 mm |
| (*c*) gap between longitudinal segment joints (packings removed and bolts tightened) | 1 mm feeler gauge not passing |

**204.4.3. Segment measuring**

1. The conformance of individual precast concrete segments to the dimensions specified on Drawings specified in Clause 204.4.1 shall be demonstrated by checks on the cast segments. The Contractor shall demonstrate the ability of each mould to produce segments conforming to the specified dimensions and tolerances on test segments to the satisfaction of the Engineer before commencing segment production. The tested segments shall be retained as reference segments until the production of segments and the associated assurance is complete.
2. For the purpose of production control, the Contractor may opt to replace some dimensional checks of segments with corresponding checks on the moulds. Measurements taken on the moulds shall be calibrated against the dimensions of the reference segments before segment production commences and periodically confirmed against measurements of segments taken from production to the satisfaction of the Engineer.
3. If either segment moulds or segments are found to be out of tolerance, all segments produced with the corresponding moulds since conformance was last established shall be quarantined until their conformance has been verified to the satisfaction of the Engineer. The mould shall also be quarantined until its dimensional accuracy and stability during use can be verified to the satisfaction of the Engineer.
4. The Contractor shall state the precision of the proposed dimensional checks and qualify the impact of external influences such as temperature.
5. The as-built joint plane shall be established by a linear regression of the offsets measured relative to a straight edge extending over the full contact area, both in longitudinal and radial direction. The grid of measurement points shall be 100 mm by 100 mm or another value accepted by the Engineer.
6. The maximum rate of deviation shall be established from the contact area survey grid points.

Figure 204.1 Acceptable tolerances for segment length and width (204.4.1 (a) and (c))

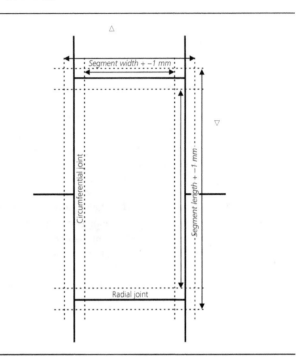

Figure 204.2 Acceptable tolerances for joint surface offset from theoretical plane in longitudinal plane (204.4.1 (j) and (k))

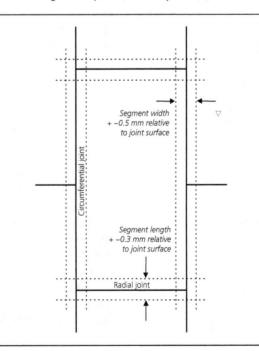

Figure 204.3 Acceptable tolerances for joint surface offset from theoretical plane in radial plane (204.4.1 (j) and (k))

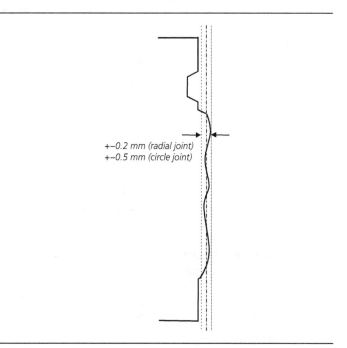

+−0.2 mm (radial joint)
+−0.5 mm (circle joint)

Figure 204.4 Combined tolerances for radial joint (204.4.1 (j) and (k))

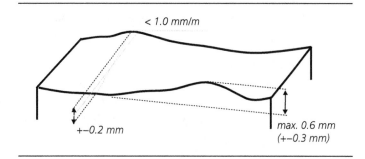

< 1.0 mm/m

+−0.2 mm

max. 0.6 mm
(+−0.3 mm)

Figure 204.5 Establishment of the theoretical joint plane from the survey grid (204.4.3.5) and rate of inclination derivation

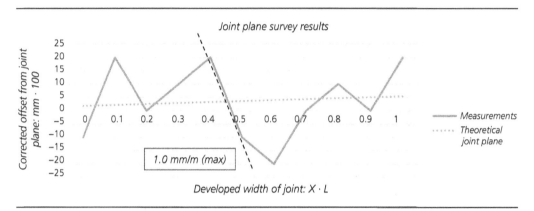

## 204.5. Tolerances for the manufacture of expanded segments

### 204.5.1. Segments

Dimensions of individual precast concrete segments shall comply with Clause 204.4.1.

### 204.5.2. Rings

Test rings shall conform with Clause 204.4.2.

## 204.6. Opening sets

Special rings and opening sets shall be built as a complete set to the tolerances specified in Clause 204.4.2 unless agreed otherwise with the Engineer. All rings of a special set shall also be built on the test ring.

## 204.7. Marking of segments

As a minimum all segments except key shall have marked with indented upper case lettering the following information on the inner face

(a) internal diameter of lining
(b) type of segment referenced to the detailed Drawings
(c) a unique mould identification
(d) any special information to indicate the position or orientation of the segment in the ring
(e) the weight of segment in kilograms
(f) date of casting or unique segment ID number.

**204.8. Identification and traceability of segments**

All major production and logistics processes such as preparation of moulds, concreting, demoulding, storage, delivery and tunnel installation shall be fully recorded. Each segment shall be identified on both the intrados surface and the circumferential joint edge using durable labels, bar codes or QR code systems. Cast-in radio frequency identification systems or similar may be used with the approval of the Engineer. The Contractor shall provide suitable scanners/readers to the Engineer and their appointed staff for any system adopted, where required.

The following data shall be captured as a minimum

(*a*) unique segment ID number
(*b*) type of segment referenced to the detailed Drawings
(*c*) date and time of casting
(*d*) type of reinforcement
(*e*) project name.

**204.9. Quality management system for segment production and delivery**

Data for all segments shall be stored in a database and made available to the Engineer. Each unique segment ID number shall refer to a dataset in the database containing

(*a*) unique reinforcement cage ID number
(*b*) type of reinforcement cage
(*c*) concrete batch information
(*d*) date and time of segment casting
(*e*) age of the segment at installation
(*f*) status of the segment (i.e., 'produced', 'in storage', 'in transit', 'delivered')
(*g*) inspection and repair history including photographic damage documentation
(*h*) location and position of the installed segment within the tunnel.

**204.10. Joint packing**

Where shown on the Drawings, stress distribution packing is to be incorporated in each longitudinal joint covering at least 80% of the intended joint contact surface area as confirmed by the Designer. The packing shall be in accordance with Clause 215.

**204.11. Gasket grooves**

Where shown on the Drawings, gasket grooves shall be provided around all joint faces of each segment and key in accordance with the dimensions recommended by the gasket manufacturer.

**204.12. Concrete cover**

1. Concrete cover shall be as stated on the Drawings or in the Particular Specification.
2. Where approved by the Designer the concrete cover to non-metallic cast-in elements can be locally reduced. The minimum cover for these elements shall be no less than maximum aggregate size +10 mm.

**204.13. Grout ports**

Where shown on the Drawings, grout ports shall be provided. The grout ports shall be designed to resist the working water pressure with a factor of safety of 2 before and after use unless specified otherwise.

**204.14. Curing**

Segments shall be cured in accordance with the provisions of BS EN 13670.

**204.15. Handling, stacking and transport**

1. The method of lifting and handling shall minimise the need for human intervention and the type of equipment and method of transport used shall ensure the security of the load and shall not damage the segments. Segments are to be stacked in a manner that ensures stability of the stack and is approved by the Engineer. The method of stacking for both storage and transport, and the acceptable positioning and size of timber battens, shall be shown on the Drawings.
2. Segments shall not be transported to site or incorporated into the works until they have achieved the 28-day compressive characteristic strength and in the case of fibre-reinforced segments their required flexural and tensile strengths.
3. If the grout port is to be used for segment handling, the Contractor shall ensure that this has been catered for in the design.

**204.16. Segments reinforced with fibres**

1. Fibre reinforcement shall comply with Clause 203.3.
2. Steel-fibre-reinforced concrete material parameters for use in segmental lining shall be determined according to fib Bulletin 83 or alternative references.
3. Guidance for the production of segments is provided in ITAtech Report 9.
4. Fibre-reinforced concrete shall be compacted and finished to ensure that fibres do not protrude from non-formed surfaces.

## 205. Spheroidal graphite cast iron (SGI) linings

**205.1. General**

1. SGI linings shall be supplied by an approved supplier who will be required to show

   (*a*) a Quality Assurance and control programme approved by the Engineer
   (*b*) a record of successful use of their linings, or an independent verification of the lining design by a competent checker in lieu
   (*c*) compliance with British Standards regarding materials, manufacture, testing and storing of materials, segments and fixings as described in BS EN 1563:2018 Founding. Spheroidal graphite cast irons.

2. The manufacturer's premises and methods shall be available for inspection by the Engineer prior to giving approval for use and at reasonable times during production for the purpose of checking the quality of manufacture. The Contractor shall ensure that all necessary assistance and testing facilities are provided to the Engineer on each visit.

3. The Contractor shall ensure that the segments are capable of being handled safely and of sustaining, without damage, forces occasioned by handling, erection and other operations.

4. Castings shall be manufactured in accordance with the Drawings and Specification. The Contractor shall produce prototype or pre-production segments to enable tests to be carried out on at least two rings of each diameter. Where the Contractor submits evidence, acceptable to the Engineer, of manufacture by their proposed source, of rings of similar size and specification, they may at their sole risk carry out the required pre-production tests on rings from initial production.

5. The materials used for castings shall have material designation EN-GJS-600-3 complying with the requirements of BS EN 1563:2018. The minimum 0.2% proof stress shall be confirmed by testing and all test results shall be supported by the appropriate documentation.

6. Castings shall be sound, clean and free from defects that in the opinion of the Engineer may affect their serviceability. They shall be properly fettled and free from sand, flashes and so on before receiving any protective coating.

7. The Designer shall specify an appropriate coating that meets all the requirements in terms of fire resistance, occupational health, environmental and smoke emissions as well as achieving the specified design life.

8. Segment lifting points shall be incorporated into the segment ribs for handling purposes. Where the segments are installed using a tunnel boring machine (TBM), the lifting points shall allow for adaption to make the SGI segment compatible with segment handling in the TBM.

9. Where the Contractor is responsible for the SGI segment design, this shall be stated in the Contract and the design requirements shall be defined in the Particular Specification.

10. Where SGI linings are used in combination with other lining types, the geometrical and material compatibility of the liners shall be reviewed by the Engineer. The Engineer shall assure the compliance of the joint system composed of SGI and other lining with the specification requirements.

## 205.2. Testing

1. The Contractor shall carry out a programme of compliance tests at times and places agreed with the Engineer. The Contractor shall provide confirmatory evidence of the test results. The Contractor shall provide the Engineer with relevant safety data sheets in advance of the tests.

2. The Contractor shall afford access and facilities to the Engineer at all reasonable times to all places engaged in the manufacture of segments to allow inspection of the production at any stage, to witness the required tests and to reject any segment that does not comply with this Specification.

3. The Contractor shall replace or rectify any segments delivered to site that are defective or do not comply with the Specification.

4. The Contractor shall carry out such additional tensile testing as may be required for proper correlation of hardness, strength and microstructure.

## 205.3. Production testing

205.3.1

The Contractor shall carry out the following tests on pre-production castings to determine the acceptance criteria

(a) spectrographic examination of each pour to determine the percentage composition of materials

(b) ultrasonic testing of all critical points on 5% of castings selected by the Engineer. The Contractor shall obtain the Engineer's agreement for the methods of testing to be employed. The degree of allowable defects shall then be agreed with the Engineer subject to the ultrasonic test results and the results of any testing described in the paragraphs below

(c) if there is any doubt as to the significance of any defects indicated by the ultrasonic tests, the Contractor shall carry out further examination by either X-ray or sectioning techniques

(d) microstructure examinations of materials from at least 5% of castings selected by the Engineer

(e) hardness tests on at least 5% of castings, including those subjected to microstructure examination

(f) additional tensile tests as may be required to establish the proper correlation of strength, hardness and microstructure

(g) where sonic or ultrasonic testing is to be used during production, preliminary tests shall be carried out as necessary to establish the mean characteristics of the material and to determine allowable deviations.

**205.3.2**   The following tests shall be carried out during production

(a) spectrographic examination shall be undertaken of each pour to determine the percentage composition of materials

(b) ultrasonic testing shall be carried out at points selected by the Engineer on 5% of all production castings. Should significant defects occur, the Engineer shall inform the Contractor to carry out ultrasonic testing on a greater number of castings. If doubt exists regarding the significance of such defects the Engineer shall inform the Contractor to carry out X-ray examinations

(c) the Contractor's supplier shall carry out hardness tests on 5% of all production castings. The Engineer shall inform the Contractor to instruct their supplier to carry out extra hardness tests on additional castings if the Engineer has doubts about the supplier's methods of manufacture or quality control

(d) castings not satisfying the stated quality standards shall be rejected, unless a programme of repairs is agreed to by the Engineer. Making good surface defects shall only be permitted where such defects are minor and then only with the Engineer's agreement. Welded repairs shall not normally be permitted

(e) the number of tensile tests shall be in accordance with BS EN 1563:2018. Sufficient samples shall be produced for testing including the extra number required in the event of a test failure

(f) if any individually tested ladle does not have a test sample made and a tensile test carried out, then at least one casting from that ladle shall have sonic/ultrasonic and hardness tests and be subject to microstructure analysis

(g) all tests shall be subject to the Engineer's agreement and may be witnessed by the Engineer or their Representative, unless specifically agreed otherwise.

**205.4. Marking segments**

1. Marks indicated on the Drawings shall be distinct and shall be cast on the inner surface of the skin of each segment or key as follows

   (*a*) internal diameter of lining
   (*b*) type of segment
       'O' ordinary
       'T' top, the 'T' shall be at the key end of the segment
       'X' special or taper
   (*c*) Employer's mark as instructed
   (*d*) mark of manufacturer
   (*e*) date of casting and mark identifying the casting with the appropriate test sample
   (*f*) weight of segment in kilograms.

2. The lettering on the skin of segments shall not be less than 50 mm high and shall project not less than 2 mm above the surface. On solid keys the lettering shall not be less than 20 mm high and may be incised.

**205.5. Machining and drilling**

1. Machining shall be carried out to a 250 mm centreline average with a grade N10 finish as defined in Table 2 of BS 1134. Machining shall be carried out before application of the protective coating system.

2. All castings shall have the radial flanges machined to correct form and dimensions as defined by the Drawings and the Specification. Where shown on the Drawings, the circumferential flanges shall also be machined in accordance with this Specification. All gasket grooves shall be machined.

3. The machined faces of segments shall normally be plane and the radial flanges shall be square to the circumferential flanges within the specified tolerances.

4. Machined surfaces shall be protected immediately after machining by the specified coating.

5. Where countersinks are required, they shall be machined concentric with the bolt holes.

**205.6. Dimensions and tolerances**

1. The accuracy of drilling bolt holes and matching flanges shall allow all similar segments to be interchangeable not only within individual rings but also with similar segments in other rings. Prior to the commencement of bulk manufacture, as a check on the casting, machining, spacing of bolt holes and

interchangeability, the Contractor shall carry out the following trials for each size of ring and taper

(*a*) assemble and bolt together on a flat level base approved by the Engineer, segments to form three rings
(*b*) the rings shall be built one above the other with the radial joints staggered by approximately half a segment
(*c*) the segments shall be bolted together with bolts 3 mm smaller in diameter than the bolt holes.

2. The lowest ring shall be maintained as a master ring for the duration of the Contract. The segments for the other two rings shall be selected at random.
3. From time to time, segments selected by the Engineer shall be built to form rings on the master rings, to ensure that tolerances and interchangeability of segments are being maintained.
4. Every taper ring shall be built on the appropriate master ring unless agreed otherwise.
5. Substantial steel templates, made in accordance with the Drawings of a design agreed with the Engineer, shall be provided, fitted with plugs 3 mm smaller in diameter than the bolt holes shown on the Drawings and of a length sufficient to pass entirely through the bolt holes.
6. Until such times as the Engineer has agreed that the setting up of the segments for machining and drilling will produce consistently accurate segments, all segments shall be built using a template.
7. When the setting up arrangements are agreed with the Engineer, sample segments shall be built using templates. The Engineer shall indicate where templates shall be used to ensure segments conform with the Drawings and this Specification. Template pins must pass freely through the bolt holes when the template inner edge corresponds with the inner edge of the flange. Template ends, both in length and angle, must correspond with the ends or sides of the segment, subject to the specified casting or machining tolerances. Master templates shall be provided for checking working templates. Working templates shall be checked every 2000 uses or every three months. The Engineer may require templates to be checked at any time.
8. Dimensions of SGI segments shall be within the following tolerances.

Table 1 SGI segment dimensions – tolerances

| Parameter | Tolerance | |
|---|---|---|
| Dimensions over a machined face | +1.0 mm | −0.0 mm |
| The thickness of any elemental part of the segment shall not deviate from the designed dimension | +3.0 mm | −0.0 mm |
| Internal diameter of a completed ring (as a percentage of the design diameter) | +0.2% | −0.2% |
| Bolt holes | | |
| Drilled diameter | +0.5 mm | −0.0 mm |
| Drill centres | ±1.0 mm | |
| Pitch circle diameter (PCD) (as a percentage of the design diameter) | +0.15% | −0.0% |
| Bolt hole for services (M10 @25mm deep) | | |
| Drilled diameter | +0.5 mm | −0.0 mm |
| Drill centres | ±1.0 mm | |
| Caulking groove | | |
| Half width dimension | ±0.5 mm | |
| Depth | ±0.5 mm | |
| Sealing groove | | |
| Depth | ±0.5 mm | |
| Width | ±0.5 mm | |
| Segment lifting point holes | | |
| Diameter | ±3.0 | |
| Deviation from pitch circle diameter (PCD) (as a percentage of the design diameter) | +0.15% | −0.0% |

Dimensions of completed master rings shall fulfil the above tolerances.

## 205.7. Segment weights

1. A schedule of computed weights for every type of segment to be supplied shall be provided by the supplier and shall be marked on the Drawings.
2. Segments weighing less than the weight computed in accordance with this Clause shall be rejected.
3. The Contractor shall make available to the Engineer copies of all delivery notes for the linings showing the weighbridge weights of each type of casting.

**205.8. Grout holes**

1. Grout holes shall be either cored or drilled, perpendicular to the internal face of the casting.
2. Unless otherwise specified, grout holes shall be threaded throughout their length in the segments and for a minimum depth of 25 mm in the inner face of solid key with 32 mm (1¼ inch British standard pipe (BSP)) parallel pipe thread. When grout plugs complying with the Specification are engaged by hand, the large end of the threaded part shall protrude from the holes by between two and four threads. An internal boss shall be provided as detailed, to give the minimum thread length detailed on the Drawings and shall not encroach on the space proofing required within the tunnel. The supplier of the SGI segments shall also supply the screw grout plugs.
3. As soon as the grout hole is tapped it shall be greased and the plug shall be screwed in from the concave side sufficiently tightly to prevent it becoming loosened or lost in transit.

**205.9. Grout plugs**

1. Grout plugs shall conform to BS 143 and BS 1256. They shall be made from malleable iron complying with BS EN 1562:2019 grade B 30-06 and shall have 32mm (1¼ inch BSP) taper heads to BS EN 10226.
2. The thread of the plugs shall be coated with grease after manufacture.
3. The segments shall be delivered to site complete with all grout plugs fitted in position.

**205.10. Casting details**

1. Washer pads, where required, shall be formed with faces perpendicular to the bolt holes.
2. Fillet radii are to be as shown per the Drawings. The Engineer will not accept sharp corners resulting from mould repairs.
3. Grommet recesses may be cored or machined.
4. Bolt holes may be cored or drilled. Cored circumferential bolt holes only shall be elongated by 5 mm.

**205.11. Corrosion protection**

1. Where corrosion protection is required, shot blasting equivalent to grade Sa 2.5 in BS EN ISO 8501-1 shall be used to fettle the castings. Immediately before protective coatings are applied, the castings shall be machined to the specified tolerances and cleaned to the original grade Sa 2.5. Machined surfaces shall be protected with a coating meeting the requirements of the Particular Specification and approved by the Engineer.
2. Grit blasting of machined faces shall be subject to the agreement of the Engineer.
3. Prior to applying the protective coating, the segments shall be pre-cleaned with water-based cleaner, thoroughly rinsed to remove all residue and allowed to dry fully.

4. The protective coating shall be applied strictly in accordance with the manufacturer's written instructions. The finished thickness of the coating shall be within the limits specified. Should this coating be removed or deteriorate during the period of storage within the control or responsibility of the Contractor or their supplier, it shall be replaced or repaired as agreed with the Engineer.

**205.12. Damaged segments**

Segments that are damaged or defective shall be indelibly marked and shall be removed from site. No damaged or defective segments shall be delivered to the Works.

**205.13. Ring removal**

Where SGI segmental rings are intended to be removed under the Contract, they shall be designed and supplied with means for safe removal.

## 206. Structural steelwork and steel linings

### 206.1. General

1. Structural steelwork shall be in accordance with the *National Structural Steelwork Specification for Building Construction* (NSSSBC) published by the BCSA and the specific requirements set out in this Specification and on the Drawings.
2. Reference to the Steelwork Contractor in the NSSSBC shall be read as a reference to the Contractor.

### 206.2. Fatigue and dynamic loading

This Specification is applicable to structural steelwork subject to static loads only. Particular requirements in respect of dynamic loads or fatigue resistance shall be specified on the Drawings or in the Particular Specification.

### 206.3. Connection design

1. For the purpose of connection design, the following details shall be as stated on the Drawings or in the Particular Specification

   (*a*) the design standards to be used for connection design
   (*b*) unfactored and factored values of the forces and their combinations at each connection
   (*c*) movements to be accommodated by each connection
   (*d*) details of the design submissions to be provided by the Contractor for acceptance.

2. Types of connection detail shall be as shown on the Drawings.
3. 'Industry standard' connection details (as noted in the NSSSBC) are not applicable.

### 206.4. Materials

1. The steel material to be used including material grade, Standard number and impact quality shall be as stated on the Drawings or in the Particular Specification.
2. Steel materials shall be tested for through-thickness properties to the specified quality class in accordance with BS EN 10164 where shown on the Drawings or stated in the Particular Specification.
3. Internal defects shall not exceed the limits set out in the Particular Specification.
4. The grades of bolt assemblies and their protective coatings shall be as stated on the Drawings or in the Particular Specification.
5. Individual components shall be traceable to their inspection and certification documents.

### 206.5. Fabrication

1. Fabrication details shall be as shown on the Drawings.
2. Thermal cutting shall not be used in areas identified on the Drawings or where stated in the Particular Specification.

3. Flame cut edges and ends shall be treated in accordance with BS EN 1090-2.
4. Welding consumables shall be such that the mechanical properties of the deposited weld metal are not less than the specified minimum values in the product standard for the parent metal being welded.
5. Where noted on the Drawings or in the Particular Specification, special welding procedures shall be submitted for acceptance prior to fabrication work commencing.
6. Particular requirements for non-destructive testing of welds, in addition to those required by the NSSSBC, shall be as shown on the Drawings or as stated in the Particular Specification.
7. Full-size punching of holes shall not be used in areas identified on the Drawings or where stated in the Particular Specification.
8. Surfaces shall be machined as stated on the Drawings.
9. Arrisses shall be smoothed by grinding or filing as necessary to allow the required thickness of protective coating to edges.
10. Flatness for full contact bearing shall be as stated in BS EN 1090-2.

## 206.6. Protective treatment

1. Grades of preparation for protective treatment shall be as stated on the Drawings or in the Particular Specification.
2. Galvanised coatings shall be applied in accordance with BS EN ISO 1461.
3. The thickness and composition of any metal coating shall be as stated on the Drawings or in the Particular Specification.
4. Post-galvanising inspection shall be as stated on the Drawings or in the Particular Specification.
5. Paint treatments shall be as stated on the Drawings or in the Particular Specification. Stripe coats shall be applied to all edges. Materials used for paint treatment of structural steelwork shall be non-flammable, shall prevent the spread of flame and shall not give off harmful gases in a fire, and the Contractor shall provide the Engineer with appropriate certificates to demonstrate compliance with these requirements

## 206.7. Steelwork erection

1. Hard stamping or other permanent identification marks shall not be used in areas identified on the Drawings as being unmarked.
2. The Contractor shall provide details of holes and attachments necessary for safety, lifting or erection to the Engineer for acceptance. Where required by the Engineer such attachments shall be removed on completion of steelwork erection.
3. Lubrication of threads for tightening of preloaded assemblies shall be in accordance with the bolt supplier's recommendations.
4. Steelwork members shall be marked, showing the weight in kilograms.

**206.8    Bolt assemblies**

1. The grades of bolt assemblies and their protective coatings shall be as stated on the Drawings or in the Particular Specification.
2. Where preloaded assemblies are required, the applicable Standard and type of system shall be as stated on the Drawings or in the Particular Specification.
3. The minimum number of clear threads protruding beyond the end of the nut and remaining between the bearing surface of the nut and unthreaded part of the shank shall be as stated in BS EN 1090-2.
4. Washers shall be used unless specified or shown on the Drawings.
5. Where bolt holes will be used for segment handling then they should be designed accordingly. Calculations shall be carried out to demonstrate that they can withstand the applied forces.
6. Where a bolt will be subjected to vibration, a suitable washer to prevent loosening shall be specified, or the bolt shall be removed where there is a risk of the bolt being displaced and where the design permits and with the approval of the Designer.

**206.9. Fabricated steel segments**

1. The procedures to be adopted for the fabrication of steel segments shall be agreed with the Engineer. Fabrication methods shall make due allowance for weld shrinkage, control of distortion, accuracy, ease of welding and avoidance of stress concentration. Preheating and stress relieving will be allowed but the Engineer may require procedural trials for the more complex joints. Templates and jigs shall be made of steel.
2. Fabrication Drawings for fabricated steel segments shall be provided to the Engineer for agreement.
3. Marking, testing, machining and drilling, dimensions, tolerances and trial rings, grout holes and grout plugs, and caulking shall follow the same general provisions as for SGI segments (refer to Clause 204 of this Specification).

**206.10. Cold-formed pressed steel segments**

1. Steel segments made by a cold-forming process (liner plates) shall be obtained from an approved manufacturer who can demonstrate

   (*a*) a satisfactory Quality Assurance and control programme
   (*b*) a record of successful production of such linings.

2. Steel used in the production of liner plates shall conform to ASTM A1011, with a minimum yield strength of 190 MPa.

**3.** Design calculations for liner plates shall be provided by the Contractor and shall prove the suitability of the chosen section in respect of

(*a*) deflection
(*b*) buckling
(*c*) stiffness
(*d*) joint strength.

**4.** Any damaged or distorted segments shall be discarded.

## 207. Jacking pipes

### 207.1. General

1. Concrete jacking pipes shall comply with the provisions of BS EN 1916, BS EN 206, BS 8500-1, BS 8500-2 and BS 5911-1.

2. Vitrified clay jacking pipes shall comply with the provisions of BS EN 295-7 and BS EN 12889.

3. Manufacturers of jacking pipes will be required to show a third-party certified Quality Assurance and control programme to ISO 9001.

4. When requested, a certificate shall be provided to the Engineer to confirm that the jacking pipes comply in all respects with the relevant standards.

5. The packing material shall be resilient and shall distribute pipe stresses arising from jacking loads. The packing material dimensions and installation shall be agreed with the Engineer prior to commencement of jacking operations.

6. The manufacturer shall provide, on request, a statement of the allowable distributed and deflected jacking loads. Details of the characteristics used in the assessment of the allowable jacking loads shall be included in the statement.

7. Provision shall be made for the injection of lubricating fluid or grout through pre-formed holes in the pipe walls. Lubrication holes shall be fitted with non-return valves.

8. All pipes shall be handled, unloaded and stacked in such a manner as to prevent damage to the pipes, in accordance with the manufacturer's recommendations.

9. Jointing shall be carried out in accordance with the manufacturer's instructions.

10. The manufacturer's premises and methods shall be open to inspection by the Engineer for the purpose of checking the quality of manufacture. The Contractor shall ensure that all necessary assistance is provided to the Engineer on each visit.

11. The planarity of the ends of the jacking pipe with respect to the pipe axis shall be defined by Table 7 of BS 5911-1, unless otherwise specified by the Engineer.

## 208. Support arches and lattice girders

### 208.1. General

1. Steel arches or lattice girders shall be installed to maintain the designed shape of the opening and, if necessary, provide an immediate support at the working face over the length of the last excavation completed. If necessary, the installation of steel arches or lattice girders shall also prevent ground loss and shall improve load distribution.
2. For the design of support arches and lattice girders the following shall be taken into account

   (*a*) axial stress and bending moment in the steel arch ribs induced by the ground loads
   (*b*) lateral stability and bracing of steel arches or lattice girders
   (*c*) method of lifting and installing the steel arches or lattice girders including the provision of lifting points
   (*d*) method of blocking and spacing of blocking points
   (*e*) bearing capacity of the ground at the toe of the arch ribs
   (*f*) the stand-up time of the unsupported part of the excavation
   (*g*) the groundwater regime and permeability of the ground.

### 208.2. Arches/ribs

1. Arches, base plates, ties and connections shall be formed from steel conforming to BS EN 10025. Arches shall be rolled to suit the dimensional requirements of the Contract. Welding shall conform with BS EN 1011-1. Holes for ties, struts and any bolted connections shall be drilled. No burning will be allowed for temporary works items or permanent elements.
2. Threaded tie rods and struts shall be of adequate length to suit arch centres and allow 25 mm projection each end beyond the nut.
3. Bolts for bolted connections shall be black bolts to BS 4190.
4. Where arches are to be provided as part of the Contractor's obligation for support the Contractor shall provide dimensional details of the arches, calculations regarding imposed loads and design and such other information that the Engineer may reasonably request.
5. Galvanised arches, where required, shall be treated in accordance with BS EN ISO 1461. All components, including the rods, fish plates, nuts and bolts shall be galvanised.
6. Where arches and ribs are used to support lagging boards they shall consist of prefabricated steel profiles. The geometry of the profiles shall be as specified by the Designer. The steel for the frames shall be S355JO or better and shall attain a minimum tensile strength of 550 MPa.

## 208.3. Lattice girders

1. Lattice girders shall consist of primary bars connected by stiffening elements to the manufacturer's design or as shown on the Drawings. They shall be designed so as to

   (a) facilitate sprayed concrete penetration into and behind the girder, thereby minimising the creation of projection shadows and/or voids
   (b) provide good-quality bonding between the steel and sprayed concrete, to form a composite structure acting as a continuous reinforced concrete lining
   (c) make allowance for the specified tolerances including convergence.

2. Stiffening elements. A minimum 5% of the total moment of inertia shall be provided by the stiffening elements. This percentage is calculated as an average along the repeatable lengths of the lattice girder. To ensure stability against buckling, the maximum spacing between the stiffening elements shall be less than three times the cross-sectional height of the girder.

3. Dimensions and tolerances. The lattice girders shall be fabricated to meet minimum clearances and tolerances shown under consideration of accuracy of placement during construction, manufacturing tolerances and lining deflection following installation. Prior to installation, each girder shall be inspected as specified below and all measurements taken shall be recorded along with any comments. Any changes in the inspection frequency must be authorised by the Designer's sprayed concrete lining (SCL) Engineer following a review of previous inspection results.

4. Each girder inspection shall check the following criteria

   (a) that the girder is fully identified with the girder type and the unique traceability reference
   (b) that the girder chord length (±25 mm) and height (±15 mm) is in accordance with the appropriate drawing detail subject to the specified tolerances
   (c) that the girder links such as straight or curved bars, 'spiders' and 'sinusoidals' are in the correct positions and are adequately welded
   (d) that the reinforcement and plate types and sizes are as specified on the Drawings.

5. When inspecting weld quality, the following criteria shall be used

   (a) the reinforcement shall be free from undercut in excess of 1 mm
   (b) the weld metal deposition shall be even and blend smoothly with the bars
   (c) the weld metal shall be free from cracks and porosity.

6. The chord length shall be checked by measuring the distance from the outer edge of the connection plate to the corresponding point on the connection plate at the other end of the girder. The measurement shall be taken to the nearest millimetre.

7. The chord height shall be taken as the measurement from the inside edge of the lower main bar at the crown to the chord formed between the outer edges of the end plates. The measurement shall be taken to the nearest millimetre. Where the girder consists of a double radius, the chord lines shall be taken along the outer edge of the connection plates to the point at which the radius changes.

8. Lattice girders shall also comply with the following tolerances

   (a) the erected lattice girders shall not deviate from the design shape and position by more than -0, +50 mm
   (b) lattice girders shall be fabricated to include an allowance for 10 mm of convergence.

9. Fabrication. Each of the primary bars of the lattice girder segment shall be composed of only one piece of either plain round profile or deformed high-yield steel to BS 4449 (minimum grade 500). Secondary bars are either plain round profile or deformed high yield to BS 4449 (minimum grade 500). In addition, all steel is to conform with Specification Series 1800 Structural Steelwork.

10. The connection elements at the end of the girder segments shall be constructed of flat or angle steel to BS EN 10025, grade S275JR. Connections between lattice girder segments shall be bolted as shown on the Drawings; welded connections between segments shall not be permitted. Nuts and bolts supplied are to be grade 8.8 or higher.

11. All welding shall be carried out in accordance with BS EN 1011-1, with welding personnel and fabrication facility UK Certification Authority for Reinforcing Steels (CARES) approved.

# 209. Spiles, dowels, rockbolts and forepoling boards

## 209.1. Spiles and canopy tubes

1. Spiles shall be either

    (a) steel bars or tubes with wall thicknesses not less than that specified and constructed from steel to BS 4449, or

    (b) glass-reinforced plastic (GRP) bars or tubes with wall thicknesses not less than that specified.

2. The spile diameter shall not be less than that specified.
3. Pre-drilled and self-drilled spiles shall be grouted. If grout is to be used for spile installation, it shall be commensurate with the ground conditions and angle of spile inclination.
4. If grout is used, specification and methods shall comply with those given in Clause 209.2.

## 209.2. Rock dowels

1. Rock dowels shall be either

    (a) untensioned steel bars threaded at one end and provided with a face plate, shim plates and a conical seated washer and nut

    (b) split or deformed steel tubes, or

    (c) glass-fibre-reinforced resin rods.

2. Steel bars shall be grade 460, deformed type 2 bars complying with BS 4449. Threaded parts of bars, nuts and seatings shall comply with the requirements of BS 4190. Face plates shall be of a dish shape in steel to the appropriate standard and shall have a hemispherical seating with centralised slot to suit dimensions of the rock dowels.
3. Where required, the bar and components shall have corrosion protection and the threaded end shall be sealed by an end cap.
4. Cement for grouting in rock dowels shall conform to the requirements of British Standards as detailed in BS 8500-2 Table 1 as appropriate to the circumstances. Cement grout shall have a water/cement ratio commensurate with the product, either thixotropic grouts or pumpable grouts and shall achieve the characteristic strength as described in Clause 304. Admixtures containing chlorides shall not be used. Other admixtures including plasticisers and expanding agents to BS EN 480 shall be used only with the Engineer's agreement.
5. Full details of resin-based grouts, including any hazards to health, shall be agreed with the Engineer. Resin grouts shall be tested in accordance with BS 6319.

## 209.3. Rockbolts

1. Rockbolts are typically passive (non-tensioned) installations. In specialist circumstances they may be active (stressed, with a debonded free length) to provide immediate support and prevent further unravelling. The bolt may be one of the following

   (a) solid steel bar (deformed) to BS 4449, or threaded bar of steel grades 500/600 N/mm$^2$ or 670/800 N/mm$^2$
   (b) hollow steel bar of the self-drilling type, grade 500/600 N/mm$^2$
   (c) slit steel tube with a tapered distal end, or folded steel tube that is expanded upon installation using high-pressure water injection
   (d) glass-fibre-reinforced resin rods, solid or hollow.

   Only item (a) can be debonded effectively for active support applications. Alternative materials shall be subject to agreement with the Engineer.
2. Where required, the bar and components shall have corrosion protection and the threaded end shall be sealed by an end cap.
3. Rockbolts shall have face plates that shall be of a dish shape in steel to the appropriate standard and shall have a hemispherical seating with centralised slot to suit the dimensions of the rockbolts.
4. Cement for grout for rockbolts where required shall conform to the requirements of British Standards as detailed in BS 8500-2 Table 1 as appropriate to the circumstances. Cement grout shall have a water/cement ratio commensurate with the product, either thixotropic grouts or pumpable grouts and shall achieve the characteristic strength as described in Clause 304. Admixtures containing chlorides shall not be used. Other admixtures including plasticisers and expanding agents to BS EN 480 shall be used only with the Engineer's agreement.
5. Full details of resin-based grouts where required shall be submitted to the Engineer for their approval. Resin grouts shall be tested in accordance with BS 6319.

## 209.4. Rock anchors

1. Rock anchors are specialised installations and are generally only required in localised areas of high load, where restraint is required, such as for stabilisation of a rock wedge.
2. Rock anchors feature a fixed length (bonded in a stable zone) and a free length (fully debonded). They are often heavily loaded and typically feature lengths of 10–30 m.
3. The anchor tendon may be

   (a) steel bar of grades 950/1050 N/mm$^2$ (prestressing steel), 670/800 N/mm$^2$ (high-strength rebar grade steel) or 500/600 N/mm$^2$ (rebar grade steel)
   (b) steel strand of grades 1770/1500 N/mm$^2$, 1820/1545 N/mm$^2$ or 1860/1600 N/mm$^2$.

4. Corrosion protection for rock anchors shall be considered in the context of design life and aggressivity of the environment. In general terms a design life of up to five years is classified as temporary. For temporary anchors an assessment of durability shall be made in line with procedures in BS EN 1537 and provided assessed corrosion does not lead to failure, no corrosion protection is necessary.

5. If design life exceeds five years or the aggressivity of the ground is deemed to present a high risk of failure, suitable corrosion protection (as outlined below) shall be provided.

6. Corrosion protection measures shall ensure the provision of a physical barrier between all areas of the stressed anchor tendon, including the head termination and the ground/environment. The integrity of the protection barrier must be comprehensive, even after installation. Particular attention shall be paid to the section of the tendon at the underside of the bearing plate that is subjected to the highest risk of corrosion.

7. Corrosion protection options include

   (a) double corrosion protection in accordance with BS 8081 – suitable for permanent works

   (b) single corrosion protection – suitable for temporary works, where additional protection to overcome local aggressivity is required. Single corrosion protection will only provide a limited degree of protection and its use should be carefully assessed by the Engineer

   (c) epoxy coating. This coating when comprehensive is highly effective; however, the coating is highly susceptible to damage and the anchors must be handled with extreme care.

8. Galvanising and sacrificial corrosion allowance only offer limited life spans in respect of corrosion protection. Furthermore, borehole grout, while beneficial where cover to the tendon is present, cannot be relied upon as a comprehensive corrosion protection mechanism as its integrity and degree of encapsulation cannot be assured.

9. Ground anchors shall be assessed and tested as prototypes and after installation according to a programme agreed with the Engineer following procedures and recommendations given in BS EN 1537.

**209.5. Forepoling and lagging boards**

1. Forepoling and lagging boards for use in tunnelling shall be steel trench sheets or special forepoling boards. The geometry and weight of the boards shall ensure they do not buckle or warp during installation, and that they provide sufficient support to the supported face. They shall be sufficiently stiff to allow installation without predrilling.

2. Where the use of forepoling and lagging boards is anticipated the Contractor shall agree the detailed construction sequence and all relevant materials with the Designer.
3. Forepoling and lagging boards shall not be acceptable as sole means of ground support.

## 210. Sprayed concrete constituent materials

**210.1. General**

1. The requirements listed below generally refer to both temporary and permanent sprayed concrete.
2. This specification is primarily for the use of wet-mix sprayed concrete but in certain circumstances dry-mix sprayed concrete may be suitable provided dust emissions are adequately controlled.
3. Sprayed concrete shall comply with Clause 202 of this Specification.
4. Minimising dust emissions at sources shall be a mix design criterion and effective control of emissions at source shall be demonstrated during spray trials.

**210.2. Cement**

Portland cement shall conform to the requirements of BS EN 197-1 or National Standards and must be suitable for sprayed concrete application.

**210.3. Pulverised fuel ash (PFA) and ground granulated blast furnace slag (GGBS)**

Pulverised fuel ash and ground granulated blast furnace slag shall conform to BS EN 450-1 and BS EN 15167, respectively, and may be included in the mix provided.

**210.4. Silica fume**

1. Silica fume shall be certified to BS EN 13263.
2. Silica fume (microsilica) shall comply with the following requirements

   (a) the content of $SiO_2$ by weight of dry mass shall be not less than 85%
   (b) the silica fume shall not contain more than 0.4% elemental silica (by weight of dry mass) or any deleterious materials such as quartz, rust and/or cellulose fibres
   (c) the specific surface area shall not be less than 15000 $m^2/kg$
   (d) the carbon content shall not exceed 2% and the total alkali content as $Na_2O$ equivalent shall not exceed 2%
   (e) $SO_3$ content (by weight of dry mass) shall be less than 2%
   (f) pH shall be $5.5 \pm 1.0$
   (g) the viscosity shall be 20 s with a 4 mm viscosity cup in accordance with British Board of Agrément Certificate 85/1568 and the relative density shall be between 1.3 and 1.4
   (h) the activity index shall be at least 100% after 28 days.

3. Testing to establish compliance with Clause 210.4.2 shall be carried out on a monthly basis.
4. Storage and handling. Silica fume in slurry form shall be regularly agitated by circulation pumps prior to use.
5. The compatibility of silica fume and liquid admixtures shall be established by carrying out appropriate accelerated testing procedures agreed with the Engineer.

**210.5. Other binders and fillers**

Other binders and fillers may be used subject to appropriate testing and trials and acceptance by the Engineer.

**210.6. Aggregates**

1. Aggregates for sprayed concrete shall comply with BS EN 12620 and Clause 202 of this Specification.
2. The maximum nominal particle size shall be 10 mm unless otherwise agreed with the Engineer. Combined gradings shall be validated through pumping and spraying trials.
3. The aggregate shall be checked for chemical reactions, such as alkali–aggregate reaction, with latent hydraulic binders and admixtures, especially accelerators.
4. The moisture content of the individual fractions of the aggregate shall be continuously monitored by moisture probes.

**210.7. Water**

Water shall comply with Clause 202.4.

**210.8. Admixtures**

1. Admixtures may be used in sprayed concrete, subject to agreement with the Engineer. Minimising the hazardous nature of admixtures shall be part of the selection criteria.
2. Admixtures shall be free from chlorides such that the percentage of chlorides shall not exceed 0.1% by weight of admixtures.
3. The required characteristic values and consistency of delivery to the site shall be agreed in writing with the manufacturer of each admixture before commencement of concrete spraying. Storage conditions and usage of admixtures shall comply with the manufacturer's recommendations.
4. Written confirmation of the stability of admixtures with the mix water shall be provided prior to commencement of site trials.
5. The content of $SO_3$ shall not exceed 4.8% by weight of total binder content.
6. Only liquid accelerators shall be used, unless pre-bagged dry mix is used where powdered accelerator has already been mixed in.
7. Only alkali-free accelerators shall be used (pH 2.0–8.0 and having alkali content less than 1% by weight $Na_2O$ equivalent).
8. Only the minimum quantity of accelerator necessary shall be permitted in normal concrete spraying operations. The quantity shall be determined by site trials, subject to maximum dosage of 8% by weight of cementitious materials. Higher dosages of accelerator can be considered subject to establishing the effect of the dosage rate on the medium- and long-term strength development on the in situ concrete. At no stage in the strength development should the strength of the accelerated mix drop below 0.7 times the strength of the unaccelerated concrete mix.

9. Testing of accelerators and the base mix with respect to acceleration of setting, early strength and decrease of strength at a later age (28 days) shall take place in due time before commencement of concrete spraying.

10. Laboratory testing of the selected type(s) of accelerator shall be carried out at dosages as recommended by the manufacturer, to establish the variability of the above properties with dosage. Accelerators showing excessive variability with dosage will not be permitted.

11. Accelerators shall be selected so that, at the dosage chosen for use in the Works, the characteristic compressive strength of any sprayed concrete at an age of 28 days can be achieved. Compliance with this clause shall be demonstrated by site trials.

12. Accelerators delivered to site shall be tested at least once every two months for their reaction with the cementitious materials used, with particular reference to the setting behaviour and strength decrease after 28 days. The stability of accelerators during storage shall be visually inspected at similar intervals. Storage times and working temperature ranges shall be in accordance with the manufacturer's recommendations. The manufacturer's safety instructions shall be observed.

13. Plasticisers and retarders complying with BS EN 934-2 may be used to reduce the quantity of the mixing water and to improve the pumpability of the concrete. The effects and optimum dosages of plasticisers and retarders shall be determined by site trials.

14. The influence of the plasticisers and retarders within the concrete mix shall be checked regularly for setting time, water reduction and development of strength. These values shall be compared with the results from the pre-commencement trials.

15. Compatibility of hydration control admixtures, plasticisers and retarders with cementitious materials and accelerators shall be verified by observation and site trials.

16. Hydration control admixtures may be used to control the hydration of the mix as appropriate to expedite construction of the Works and reduce wastage. The effects and optimum dosages of hydration control admixtures shall be determined by site trials.

17. Hydration control admixtures shall be used in accordance with the manufacturer's instructions.

## 210.9. Consistency

1. Flow shall be determined in accordance with BS EN 12350-5. The flow range should be set to ensure that the pump filling efficiency is greater than 80% at all times. The flow range for a mix should be set during the trial mix development. Generally, flow values of between 50 cm and 68 cm will give acceptable performance.

2. The temperature of the plastic concrete should be between 10°C and 35°C at all times during batching delivery and application. The temperature range may be extended if proven by site trials and accepted by the Engineer.

## 210.10. Strength and quality

1. The compressive strength of the sprayed concrete in the short and long term shall be specified by the Designer. Where early-age support is crucial to support unstable ground or minimise ground movements, higher early-age strength will need to be specified.
2. The compressive strength of sprayed concrete at the age of 28 days shall be in accordance with BS EN 206.
3. The minimum early-strength development shall conform to Table 2, unless otherwise specified by the Designer.

Table 2 Sprayed concrete early strength development

| Age | In situ strength: MPa |
| --- | --- |
| 1 h | 0.5 |
| 3 h | 1.0 |
| 9 h | 2.0 |
| 12 h | 2.5 |
| 24 h | 5.0 |
| 28 days | 27.2* |

* For a C32/40 mix, with the reduction factor of 0.85 for cores from in situ concrete as per BS EN 13791 Table 1

4. For permanent sprayed concrete the coefficient of water permeability shall be less than $1 \times 10^{-11}$ m/s at 28 days according to the test method described in Concrete Society Technical Report 31 *Permeability of site concrete*, or the penetration depth should be less than 50 mm at 28 days according to BS EN 12390-8, unless otherwise specified by the Designer or agreed with the Engineer.

## 210.11. Fibres

1. The required structural performance of fibre-reinforced sprayed concrete shall be established by the Designer. The Designer shall specify the testing required to meet the design criteria.
2. Fibre reinforcement shall comply with Clause 203.3.

**210.12. Regulating layer**

1. Where used, regulating layers shall consist of sprayed concrete and shall use an aggregate not exceeding 4 mm.

2. Any regulating layer to be applied before application of a sheet waterproofing membrane shall meet the requirements of the waterproofing system as set out in Clause 211 of this Specification.

3. A regulating layer to be applied as the final finish for any sprayed concrete secondary lining shall cover all fibres and provide a closed surface that will not hold water when cleaned.

## 211. Sheet waterproofing membranes

**211.1 Sheet waterproofing membrane systems**

1. Sheet waterproofing membrane systems for tunnels shall comprise a drainage and protection layer fixed to the primary lining substrate, with a sheet waterproofing membrane fastened to this; see Clause 312 for details of installation.
2. Where the tunnel is situated above groundwater level, the groundwater seepage shall be collected as far as possible at the discharge point in the tunnel reveal and shall be fed by way of a primary drainage system to the secondary system drainage (longitudinal tunnel drainage).
3. Where the tunnel is situated below groundwater level and the water table cannot be permanently lowered, the cross-section and waterproofing system shall be dimensioned for a design under water pressure.

**211.2 Materials – drainage and protection layer**

1. The sheet membrane shall be protected against mechanical puncture by a geotextile fleece or geocomposite drainage layer, which also provides a drainage path for any water seepage around the tunnel structure.
2. The drainage and protection layer will also create a sliding surface to minimise tension and stress forming in the membrane and allow dissipation of the stresses, such as those generated through early-age thermal behaviour and settlement, generated in a secondary lining.
3. Geotextile fleece shall be a non-woven fleece and conform to the performance requirements shown in Table 3A.
4. Geocomposite drainage layer shall be a studded drainage membrane made from thermoplastic material and conform to the performance requirements shown in Table 3B.
5. The drainage and protection layer is to provide adequate mechanical protection and protection from chemical aggression caused in the curing processes of concrete.
6. Water transmissivity of the drainage and protection layer should be designed to suit expected volume of water ingress.
7. The waterproofing substrate and drainage and protection layer shall be coordinated to ensure hydraulic efficiency to discharge any seepage water present into the secondary drainage system at minimal pressure. Where this cannot be demonstrated, additional surface drainage elements with enhanced drainage capability shall be installed.

Table 3A Performance requirements for geotextile fleece

| Property | Test method | Requirement |
|---|---|---|
| Mass per unit area | BS EN ISO 9864 | Not less than 800 g/m$^2$ (invert) Not less than 500 g/m$^2$ (elsewhere) |
| Thickness at specified pressure | | |
| 2 kPa | BS EN ISO 9863-1 (Method A) | As specified by manufacturer |
| 200 kPa | | 800 g/m$^2$: $\geq$ 3.4 mm 500 g/m$^2$: $\geq$ 1.7 mm |
| Fire rating | BS EN 13501-1 | Class E |
| Static puncture | BS EN ISO 12236 | 800 g/m$^2$: $\geq$ 9000 N 500 g/m$^2$: $\geq$ 5500 N |
| Water permeability in geotextile plane at 200 kPa at HG1 | BS EN ISO 12958 | $\geq$ 0.003 l/(m · sec) |

Table 3B Performance requirements for geocomposite drainage layer

| Property | Test method | Requirement |
|---|---|---|
| Mass per unit area | BS EN ISO 9864 | To be agreed with the Engineer |
| Thickness at specified pressure | | |
| 2 kPa | BS EN ISO 9863-1 (Method A) | As specified by the manufacturer |
| 200 kPa | | Not less than 3.4 mm |
| Fire rating | BS EN 13501-1 | Class E |
| Static puncture | BS EN ISO 12236 | Not less than 500 N |
| Water permeability in geotextile plane at 200 kPa at HG1 | BS EN ISO 12958 | $\geq$ 0.5 l/(m · sec) |
| SIM creep at 200 kPa | ASTM D7361 | Not more than 20% at 100 years |

**211.3. Materials – fixing elements for sheet waterproofing membrane**

1. The drainage and protection layer is fixed onto the substrate with non-projecting disks. The disks are secured through the geotextile and into the substrate with shot-fired nails.
2. The disks should be made of a compound that allows the sheet waterproofing membrane to be fully welded to the surface.
3. In order to prevent stresses being transferred from the secondary lining to the sheet waterproofing membrane, the resistance to failure in shear of the nails and disks must be less than the shear resistance of the sheet membrane itself.

**211.4. Materials – sheet waterproofing membrane**

1. The sheet waterproofing membrane shall consist of a continuous impermeable heat-welded sheet of one of the following materials

   (*a*) soft polyvinyl chloride (PVC) unreinforced
   (*b*) flexible polyolefin (FPO/TPO) unreinforced.

2. The membrane as supplied shall be of such dimensions and shape as will result in the minimum of on-site seam welds.
3. Unless otherwise stated in the Contract, the membrane shall conform to performance requirements and have properties shown in Table 4.
4. Further information on test methods and requirements for mechanical properties and durability can be found in the OEBV Guideline for Tunnel Waterproofing.

Table 4 Performance requirements of sheet waterproofing membranes

| Property | Test method | Requirement |
| --- | --- | --- |
| Thickness | BS EN 1849-2 | 2.0 mm ±10% |
| Tensile strength | BS EN ISO 527-3 | 16 MPa |
| Elongation at break | BS EN ISO 527-3 | Not less than 300% |
| Resistance under water pressure | BS EN 1928 (Method B) | 5 bars at 1 h |
| Root resistance where required in the Particular Specification | PD CEN/TS 14416 | No penetration |
| Tensile strength of welded seam | BS EN 12317-2 | Cracks occur next to the seam |
| Water absorption | BS EN ISO 62 | <4.0% |
| Fire rating | BS EN ISO 11925-2 | Class E |

5. Where concrete is to be placed against the sheet waterproofing membrane, a signalling layer, to give a visual indication of any mechanical damage, shall be provided on the exposed surface of the waterproofing membrane. The signalling layer shall be such that it does not adversely affect the seam welds.

6. The membrane shall not be used for transfer of shear forces and shall be capable of absorbing potential movements in adjacent structural parts without damage.

7. The selected membrane system shall be suitable for the expected stresses.

### 211.5. Additional items – sheet waterproofing systems

1. The waterproofing system shall be divided into sectors with the water bars formed of material welded to the sheet waterproofing membrane. A sector shall be no more than 12 m in chainage.

2. Additional drainage capacity can be provided by studded drainage membrane made from thermoplastic material attached prior to installation of the geotextile fleece.

3. Double-sleeved reinjectable hoses with offset openings and/or slots to dispense compressed injection material can be used to seal joints and fill the cavity formed during the casting of the secondary lining. The hose should be made of a material compatible with attachment to the sheet waterproofing membrane. The openings in the interior hose are to be offset from the openings in the outer hose to prevent the entry of any injection material.

4. If reinforcement is required for constructions with sheet waterproofing membranes, it shall generally be self-supporting. Where this is not possible, anchors shall be approved for the membrane system by the membrane manufacturer.

5. Where items such as water bars are welded to the sheet membrane, they shall be of a material compatible to the sheet membrane. The compatibility shall be confirmed by the manufacturer of the membrane and the item.

### 211.6. Requirements for the waterproofing system substrate

1. The substrate for the waterproofing system shall have adequate dimensional stability and resistance and meet surface appearance and compressive strength requirements defined for the membrane system and any necessary protective layers and surface drainage elements.

2. The material properties of the waterproof system substrate and the fixings for the membrane shall be compatible. The fixings shall not loosen from the substrate until the inner shell is completed. The substrate must not spall when nails are shot-fired.

3. Water flowing through the substrate shall be collected and drained before commencing the installation of the waterproofing system.

## 212. Spray-applied waterproofing membrane

### 212.1. General

1. Spray-applied waterproofing systems include waterproof linings formed in situ and cured in place.
2. Selected spray membrane systems must permit the safe construction of the secondary lining (cast in situ or sprayed) without reduction in waterproofing properties.

### 212.2. Materials

1. The materials will be prepared in accordance with the manufacturer's instructions. No site batching variations from these instructions will be permitted without written agreement from all parties including Engineer, Designer and Manufacturer.
2. The product shall conform to the performance requirements shown in Table 5.

Table 5 Material performance criteria

| Property | Requirements |
| --- | --- |
| Bond to substrate | Failure shown to be in substrate or bond >0.5 MPa |
| Permeability | Zero penetration of water through membrane |
| Crack bridging | Capable of bridging a 2 mm gap without diminishment of resistance to water permeation |

3. Storage conditions of the product shall comply with the manufacturer's recommendations.
4. The manufacturer shall demonstrate the durability of the product for the design life of the project.

### 212.3. Materials – health and safety during application

1. The Contractor shall supply the Engineer with safety data sheets for all spray-applied waterproofing materials intended to be used. Where a product presents a fire-related hazard or is considered a hazardous material in terms of the Control of Substances Hazardous to Health Regulations (COSHH) and no less hazardous substitute is reasonably available, then measures shall be put in place to control these hazards. This requirement applies to both storage and application of the product. Appropriate measures regarding storage shall include

(a) adequate ventilation
(b) fire detection
(c) fire suppression
(d) secure storage
(e) specialised personal protective equipment
(f) inclusion of appropriate information in the site emergency procedures available to first responders

(g) product-specific emergency procedures agreed in advance with emergency services

(h) information on location of storage and quantity stored available to first responders

(i) appropriate hazard warning signage at storage locations.

Measures regarding application shall include

(a) remote application

(b) trained operatives

(c) adequate ventilation

(d) eye wash and first aid facilities

(e) exclusion zones

(f) specialised personal protective equipment.

2. For a particular product, where some or all of the identified hazards are adequately covered by a manufacturer's recommended application method, then when this product is used it shall be applied strictly in accordance with those recommendations.

3. Procedures shall be developed for dealing with spillage of materials. The procedures shall take due account of the toxic nature of the substances concerned.

**212.4. Selection of spray-applied membrane systems – from track record or appropriate trials**

1. The capability of the equipment, workmanship, materials and application methods under field conditions should be demonstrated by either

(a) previous relevant performance in similar conditions for projects with equivalent acceptance criteria

(b) appropriate trials.

2. The criteria for the acceptance of the applied waterproofing membrane's performance shall be in accordance with the project-specific requirements for degree of watertightness (please refer to Clause 508 – or BS 8102 Code of practice for protection of below ground structures against water from the ground).

3. The testing programme shall be started sufficiently early and prior to installing the membrane to allow verification that the required watertightness can be achieved and allow repetition of the trials should the initial results prove unsatisfactory. All trials, acceptance tests and development of a safe system of work shall be completed satisfactorily by the time installation commences.

4. Evidence must be available to demonstrate safe application of the proposed secondary lining within the chosen spray-applied waterproofing system. For a sprayed concrete secondary lining this evidence shall demonstrate that a sprayed lining can be applied to a fully cured spray membrane test section in the main tunnel crown with no observed instability of the freshly applied sprayed concrete, such as sagging or sprayed concrete fallout.

## 212.5. Quality Assurance and requirements during trials

1. Trials, as required by the Engineer to validate previous data, shall be carried out within the tunnel to assess the performance of the spray-applied waterproofing system in all conditions to be encountered during the permanent works and to evaluate how many layers of sprayed membrane are required to achieve the dry film thickness, including where appropriate

   (*a*) a dry area
   (*b*) a damp area
   (*c*) an area with active groundwater ingress.

2. The waterproofing system can include integrated water management measures as appropriate, including

   (*a*) locally applied grout/injection systems
   (*b*) faster curing mix solutions
   (*c*) active drainage such as strips or pipes.

3. The test area shall be sufficiently large to adequately represent the permanent situation, and to evaluate how many layers of sprayed membrane are required to achieve the dry film thickness by overlap >300mm between layers.

4. During trials the membrane and waterproofing shall be applied using the same equipment and methods, and by the same approved personnel, as those intended for the permanent works.

5. Substrate surface roughness – the trials shall be carried out on the full range of surface roughness to be encountered during application of the permanent works. This trial shall confirm the requirement or otherwise for smoothing layers additional to those required as part of the standard waterproofing system.

6. Trials shall determine the factors needed for successful application, including

   (*a*) the number of layers required to achieve the dry film thickness specified
   (*b*) the overlap between layers required

(c) the use of different colour dyes to ensure coverage of each layer

(d) training of operatives to control application to prevent sloughing

(e) curing time against layer thickness to achieve the Shore hardness required before spraying concrete onto the membrane

(f) methods of repairing defects.

7. A visual inspection of the spray-applied waterproofing membrane shall be carried out. Areas in which the substrate is still visible, the spray-applied membrane is not sufficiently opaque (for opaque coloured membranes) or where the spray-applied waterproofing membrane is damaged, shall be marked up and an additional layer of spray-applied waterproofing membrane applied with a minimum lap of 200mm around the area.

8. Where the trials are carried out in the tunnel the spray-applied waterproofing membrane shall be repaired as for a defect, as detailed in Clause 212.5.6.

9. Manufacturers shall detail post-application non-destructive testing to identify the integrity (in accordance with the specified dryness criteria) of the lining. These tests shall be carried out at a frequency and spacing as identified by the manufacturer. Where the integrity of the lining is shown to be insufficient, repairs shall be carried out in accordance with Clause 212.5.6.

10. In order to provide an additional reference during full-scale application, the quantity of spray-applied waterproofing membrane applied to achieve the required thickness per metre squared (over the given substrate condition) shall be assessed and recorded.

11. Cores (concrete–membrane–concrete sandwich) and patches shall be taken from test panels or the testing area as required by the specification, in order to demonstrate the properties of the combined system specified in Table 6.

12. The trials shall include measurements of atmospheric contamination produced during the spraying process in the form of particulates and gaseous contaminants. These shall be used to develop a safe system of work for the spraying process and, in particular, to ensure exposure to contamination by those affected by the spraying process is kept within current exposure limits in HSE publication EH40. The objective shall be that no one other than the spraying operative should require to wear respiratory protective equipment when undertaking their normal functions in the vicinity of the spraying process.

Table 6 Trial criteria for spray-applied waterproofing membranes

| Property | Test method | Requirements |
| --- | --- | --- |
| Bond to substrate | BS EN ISO 4624:2016 – Pull-off test for adhesion, for resin-based materials (using a 50 mm dolly) or BS EN 1542:1999 – Pull-off test, for all other materials. Test dollies fully stored in water for 28 days. | Failure of the substrate or bond >0.5 MPa at 28 days (as evidence of long-term water path obstruction) |
| Permeability | BS EN 12390-8 (but sealed and tested for 28 days with spray-applied waterproofing membrane located <25 mm from tested face of the specimen, within the primary and secondary layers) or Taywood Testing/similar appropriate where the lining is put to a 10 bar test for 28 days | Zero penetration of water through membrane |
| Crack bridging static test | BS EN 1062-7:2004 Part 7 | Capable of bridging gap width that is the minimum of <br><br> a) planned gap width (e.g. at joints) <br> b) 2 mm <br><br> without diminishment of resistance to water permeation. |

**212.6. Quality Assurance construction testing**

1. Coverage/continuity. A visual inspection of the spray-applied waterproofing membrane shall be carried out. In addition, manufacturers shall detail post-application non-destructive testing to identify the integrity of the lining. These tests shall be carried out as specified in Table 7 below. Areas in which the substrate is still visible, or where the spray-applied

waterproofing membrane's integrity is impaired, shall be
marked up and an additional layer of spray-applied
waterproofing membrane applied with a minimum lap of
200 mm around the area.

2. **Thickness.** Product allowing, thickness measurements shall
be carried out, as specified in Table 7, using a simple depth
gauge. The equipment used shall be approved by the Engineer,
with the thickness and location of the test recorded as
appropriate. Adequate applied thickness may be additionally
cross-referenced to the quantities per metre squared identified
during trials.

3. The location of the thickness and coverage tests shall be
determined to give even distribution around the entire lining –
that is, samples from the crown, axis and invert.

Table 7 Construction testing for spray-applied waterproofing membranes

| Parameter | Test method | Frequency | Pass/fail criteria |
|---|---|---|---|
| Coverage/ continuity | Visual | A visual inspection to be carried out continuously while the membrane is applied | 100% coverage, Where appropriate: Lining should be 100% opaque |
| Thickness | Wet film thickness – depth gauge | As required in the Particular Specification but minimum 10 tests per 100 m$^2$ | As per manufacturer's recommendations in given conditions, verified by site trials |
| | Application quantity measurement | Per batch | kg/m$^2$ to match minimum applied quantity determined during field trials |
| | Patch test | As required in the Particular Specification but minimum 1 test per 200 m$^2$ | As per manufacturer's recommendations in given conditions, verified by site trials |

## 213. Gaskets

### 213.1. Compression gaskets – general

1. Gaskets for precast concrete segmental lining shall be supplied by a specialist supplier certified to ISO 9001 or equivalent quality standard.

2. Unless specified or detailed otherwise, the recommendations and requirements of document *Recommendation for gasket frames in segmental tunnel linings* published by STUVA (Studiengesellschaft für Tunnel und Verkehrsanlagen e.V.) shall be followed.

3. Gaskets shall provide a seal against the ingress of groundwater during construction and in the permanent condition. Gasket material shall comply with the requirements of BS EN 681-1 Elastomeric seals. Material requirements for pipe joint seals used in water and drainage applications – Vulcanized rubber, or similar approved the Engineer.

4. The gasket material shall withstand any aggressive response from the ground or groundwater conditions to be encountered including the medium contained in the tunnel. In particular, the gasket material shall withstand chemical attack and biological/microbial degradation such that the gasket functions properly for the required design life.

5. Where required by the project, the gasket material shall comply with the requirements of BS EN 682 Elastomeric seals. Materials requirements for seals used in pipes and fittings carrying gas and hydrocarbon fluids.

6. The extruded gasket section shall be joined to form a gasket 'frame' to suit the individual segment dimensions and corner angles. For a glued-in gasket this shall be a stretch fit into the grooves of the concrete segments. For a cast-in gasket the frame shall be dimensioned to ensure a secure and stable fit in the mould with no grout loss around the seal with the mould during concrete placement, vibration and curing, and no damage to the concrete or gasket during demoulding.

7. The corner joint shall be a vulcanised joint and the corner pieces shall be of a different section from the extruded lengths in order that the watertightness characteristics described in this specification may be achieved and to avoid excessive load on the corners of the concrete segments. Other joints, for example for compartmentalising gaskets where double gasket systems are used, shall be acceptable where detailed on the Drawings or agreed with the Engineer.

8. Gaskets shall be fixed to the segmental tunnel linings prior to erection. The adhesive for glued gaskets shall be as recommended by the manufacturer of the gasket.

9. Gasket faces shall be lubricated prior to erection with a product demonstrated not to affect the long-term performance of the

gasket. The lubricant shall be recommended by the gasket manufacturer and agreed with the Engineer. The Contractor may propose a product and/or an alternative method of installation without lubrication, as long as the long-term performance of the gasket is not affected, for agreement with the Engineer and gasket manufacturer.

## 213.2. Compression gaskets – testing

1. Gaskets shall be tested in accordance with the STUVA document *Recommendation for gasket frames in segmental tunnel linings*.

2. The test rig for assessing watertightness shall simulate a range of conditions of displacement and joint gap, including the worst tolerance and packing combination to be encountered in the completed structure, and the type of joints to be constructed in the tunnel. A test rig utilising curved samples to eliminate the need for mitred corner joints may be used although particular care will need to be taken with the gasket height for tight radii.

3. The watertightness test pressure shall take into account the relaxation behaviour of the gasket and a factor of safety. Unless specified otherwise the test pressure shall be calculated based upon

$$\text{Test pressure} = \frac{\text{Maximum working pressure} \times \text{FOS}}{\text{Residual stress following relaxation}}$$

The residual stress following relaxation shall be taken as (100 − stress relaxation (%)) / 100
The factor of safety (FOS) shall consider other uncertainties such as the variation of the water pressure, the effect of air entrapment in the gasket voids during the test and the use of different test rig materials. A minimum FOS of 1.3 is recommended.
A minimum test pressure of 2 times the working pressure should be applied unless specified otherwise.

4. The load–displacement behaviour testing of the gasket shall be undertaken to replicate conditions equivalent to those anticipated on site, and in particular the following shall be considered in addition to the requirements of the STUVA document *Recommendation for gasket frames in segmental tunnel linings*

   (*a*) the ends of test pieces shall be sealed to prevent air loss from the voids during testing to represent a fully sealed gasket frame. The seal shall be formed using a thin layer of similar material to the gasket and shall minimise effects at the ends where a seal against the closed ends of the form cannot be demonstrated

(b) the rate of loading during the test shall be representative of the movement speed of the segment erector where such a device is used

(c) results shall include load–displacement results for the worst tolerance combination of the gasket and segment

(d) evidence shall be provided that the results for a range of gasket offsets are no worse than the test without offset, or results provided for the worst case.

5. All load–deflection, relaxation, restorative, spalling and watertightness test pieces shall be cut open and examined for visual inspection. The profile shall be checked against the Drawings and tolerances to ensure that it conforms with requirements. One piece of the samples tested in each test type should be retained by the Contractor or supplier of the gaskets until the end of the project maintenance period.

6. Testing undertaken for previous projects following the requirements specified within this document may be accepted with the approval of the Engineer. Spalling behaviour should consider, in particular, the specific requirements identified in the STUVA document *Recommendation for gasket frames in segmental tunnel linings*, including the anchor detail for anchored gaskets. Spalling tests shall consider worst case internal and external gasket manufacturing tolerances and segment tolerances.

**213.3. Hydrophilic gaskets**

1. Hydrophilic sealing material shall perform to the same effect as compression gaskets. The composition and properties of the proposed material shall be agreed with the Engineer and sealing strips and joints shall be subjected to the same testing regime as set out for compression gaskets. A time allowance for the expansion of the hydrophilic gasket shall be allowed for in the test where appropriate.

2. Hydrophilic gasket material shall take into account the ground and groundwater conditions to be encountered including the medium contained in the tunnel and shall withstand chemical attack and biological/microbial degradation such that the gasket functions properly for the required design life.

3. Hydrophilic gaskets shall be of an extruded hydrophilic rubber of an appropriate profile and size to fit preformed grooves in concrete segments. The gasket shall be treated with a coating to delay the onset of swelling during erection of segments.

4. Hydrophilic gaskets shall be protected from contact with water, including rainwater, prior to erection.

**213.4. Composite gaskets** Composite compression and hydrophilic gaskets shall meet the requirements of the relevant clauses above and shall be tested to the same requirements as compression gaskets. A time allowance for the expansion of the hydrophilic portion of the gasket shall be allowed for in the test where appropriate.

**213.5. Gaskets for pipe jack joints**

1. Gaskets for pipe jack joints shall provide a seal against the ingress of groundwater during jacking and in the permanent condition. Gasket material shall comply with the requirements of BS EN 681-1, including resistance to chemical attack and microbiological degradation.
2. The gasket shall be lubricated with a product recommended by the manufacturer and agreed with the Engineer.

## 214. Cementitious grout

### 214.1. General

1. Setting times and strength of grout shall be proposed by the Contractor for the acceptance of the Engineer, or as proposed on the Drawings.
2. General-purpose cement grout shall be mixed in accordance with the proportions given in Table 8. The water content shall be kept to the minimum required to ensure a smooth, fluid mix.
3. Pulverised fuel ash (PFA) shall not be used as a constituent of grouts that contain sulphate-resisting cement.
4. Cement grout shall be used within one hour of mixing unless its use at a higher age is demonstrated by specific testing to the satisfaction of the Engineer.
5. Grout for contact grouting of cast-in-place concrete secondary linings shall be non-shrink, suitable for completely filling the void and have a final characteristic strength equal to the 28-day characteristic strength of the concrete to be filled.
6. Grout setting times and strengths shall be agreed with the Engineer.
7. Details of admixtures, such as accelerating and retarding agents for proposed inclusion within the grout mix, including safety data sheets shall be submitted to the Engineer for agreement.
8. Preconstruction grout trials shall be undertaken to demonstrate that the required setting times and strengths will be achieved. Details of the trials and results shall be submitted to the Engineer for approval.

Table 8 Mix proportions for cement grout

| Class | Proportion by mass | | |
| --- | --- | --- | --- |
| | Cement | Sand | PFA |
| G1 | 1 | – | – |
| G2 | 1 | 3 | – |
| G3 | 1 | 10 | – |
| G4 | 1 | – | 10 |
| G5 | 1 | – | 4 |
| G6 | 1 | – | 0.5 |

9. Records of batching and batcher calibration shall be maintained to demonstrate that grout batching is in accordance with the design mix. Alternatively, grout strength tests may be used.

10. Robustness of the mixes shall be demonstrated to the satisfaction of the Engineer by varying the water-cement ratio and density of the grout by 10% or another value agreed with the Engineer, and additives within the stated precision of the dosage equipment.

11. Re-use of grout or combination of fresh grout with pre-produced grout shall not exceed a proportion of 10:1 unless demonstrated by testing to the satisfaction of the Engineer.

## 214.2. Special grouts

1. Primary grout for machine-driven tunnels shall be special grout according to this Clause.

2. The Contractor shall propose the grout mix for the acceptance of the Engineer. The Contractor shall demonstrate that the grout

   (a) ensures full contact of the gap between the excavated diameter and the primary lining extrados for the full design life when considering grout shrinkage

   (b) provides adequate support to the segment ring while curing, limiting deformations as the ring leaves the shield, and as early age loads such as trailer loads are applied

   (c) provides adequate support to the segment ring by compliance with mechanical parameters specified by the Designer

   (d) does not contaminate the ground or groundwater.

3. Clause 202 applies for all constituents of special grouts. The grout mix including all materials for proposed inclusion shall be submitted to the Engineer for agreement including safety data sheets.

4. Unless confirmed by the Designer, the Contractor shall select the grout properties so that a higher stiffness than the surrounding ground is achieved at the time of first loading.

5. Preconstruction grout trials using the materials intended for use in situ shall be undertaken to demonstrate that the required properties will be achieved. Details of the trials and results shall be submitted to the Engineer. The preconstruction trials shall replicate the planned grouting process, so far as possible, and shall be undertaken for a temperature range representative of the conditions anticipated for site use.

6. The preconstruction trials shall as a minimum capture the following parameters

Table 9 Preconstruction trial parameters

| Parameter | Test | Comment |
|---|---|---|
| Temperature | EN ISO 1 | For all tests |
| Density | BS EN 12350-6 | |
| Viscosity | ASTM D6190 | Bi-component: test grout Single phase: test bentonite |
| Bleed | ASTM C940 | |
| Gel time | Bucket test | Bi-component only |
| Compression modulus | BS EN ISO 17892-5 | At 3h/6h/24h/7d/28d |
| Unconfined compressive strength | BS EN ISO 17892-7 or BS EN 12390-3 | At 3h/6h/24h/7d/28d |
| Modulus of elasticity | BS EN 13412 | At 24h/7d/28d |
| Shear strength | BS EN ISO 22476-9 | Adopt to expected shear At 3h/6h/24h/7d/28d |

7. The Contractor shall propose details of the primary grout, including the required setting times and strength and stiffness gain to prevent ring distortion and where necessary support the weight of the tunnel boring machine (TBM) and the backup. As a minimum the initial set of the grout shall be achieved within 45 min of injection at 20°C, and the minimum strength requirement of the grout as measured from testing 100 mm cubes shall be 1.5 N/mm$^2$ in 24 h, unless the adequacy of a different setting time and/or grout strength has been demonstrated by the Contractor to the satisfaction of the Engineer. The proposals shall be submitted to the Engineer for agreement prior to commencement of the Works.

8. Records of batching and batcher calibration shall be maintained to demonstrate that grout batching is in accordance with the design mix. Production control of grout shall further include as a minimum compressive strength testing.

9. The utilisation of special grout in compression shall be limited to 50% of the characteristic unconfined compressive strength observed in tests unless a higher utilisation can be maintained in creep tests linking to the specific limit state. A test duration of 28 days shall normally be considered adequate for permanent design situations.

10. Grouting equipment shall be fitted with a pressure gauge and automatic pressure release valves capable of being pre-set to a specific pressure. Grout pressure is to be measured at the nozzle with a suitable gauge. Alternatively, grout pressure can be controlled by pressure sensors and pre-set to a designed grouting pressure for termination of the grouting process.

## 214.3. Mixing

1. Grouts containing polymer additives shall only be mixed in a colloidal-type mixer.
2. Special grouts from proprietary suppliers shall be mixed and used in accordance with the supplier's instructions.

## 214.4. Storage and delivery

1. Bagged grouts shall be stored under cover in dry surroundings and on a suitable platform, clear of the ground.
2. Bulk deliveries of grout constituents shall be stored in appropriate silos with suitable dust control and batch weighing equipment.

## 215. Packings

### 215.1. Packings for segmental linings

1. All forms of packing shall be of a shape commensurate with the contact area, provided with bolt holes where required and of a width that does not prevent the proper operation of any gasket or seal included in the joint. All packer materials shall be as detailed on the Drawings or approved by the Engineer.
2. Timber packings shall be knot-free softwood, plywood or fibreboard, sawn to shape with bolt holes where applicable. They shall be treated to retard rot and fire and shall be available in all necessary thicknesses.
3. Synthetic packings shall be cut to shape with bolt holes where applicable. The nominal peak service load of the packers and the E modulus for all provided thicknesses shall be specified by the supplier. Where no relevant performance data or test data for the general performance of the material is available, the Poisson ratio and friction coefficient of the packer material shall be specified for a test load of up to 1.2 times the nominal peak service load.
4. Where packings are subjected to planned cyclic loading the peak service load and its associated deformation shall be demonstrated for a number of load cycles agreed with the Engineer.
5. Packings shall only be used where detailed on the Drawings or agreed with the Engineer.

### 215.2. Packings for opening frames

1. Packings and folding wedges for opening frames shall be as detailed on the Drawings.
2. To prevent buckling, all packings and folding wedges shall be located at circumferential joints of the tunnel lining segments and at stiffeners in opening frames.

### 215.3. Packings for jacking pipes

Packings for jacking pipes are included in Clause 207.

### 215.4. Packings for timber headings

Timber wedges and packings shall be as detailed on the Drawings.

### 215.5. Fixings for packers

1. Glues used for attaching packings shall ensure the secure attachment of the packer for all environmental conditions encountered post attachment.
2. Mechanical fixings are only permitted if designed not to interfere with the performance of the packer. Mechanical fixings shall only be used where agreed with the Engineer.

## 216. Grommets, bolts, dowels

**216.1. Grommets**

1. Grommets for precast concrete linings shall be polyethylene.
2. Grommets for SGI, cast iron or steel linings shall be either of low-density polyethylene or gel grommets.

**216.2. Bolts**

1. Bolts shall generally be black bolts to BS 4190.
2. Sherardised bolts, where required, shall be treated to BS 7371-8 Class 75. A lower class can be selected subject to agreement with the Engineer.
3. Galvanised bolts shall be hot-dip spun to BS EN ISO 1461.
4. Stainless steel bolts, where required, shall be to BS EN ISO 3506-2.

**216.3. Dowels**

1. Dowels in the context of this Clause are any elements connecting rings of segmental linings in the circumferential joint plane. Dowels can transfer shear and/or tension force, as specified by the manufacturer.
2. Dowels, where used, shall be agreed with the Designer and specified on the Drawings.
3. The mechanical properties of dowels shall be specified by the dowel supplier and shall be considered in the design. The dowel supplier shall confirm that all production tolerances for their product are reflected in their specification and shall provide upper bound and lower bound values for mechanical parameters where required. As a minimum, the following mechanical properties shall be specified

    (*a*) nominal peak tensile load and associated tensile deformation
    (*b*) nominal peak shear load and associated shear deformation
    (*c*) ratio of short-term to long-term performance
    (*d*) gap before engagement of the assembly.

4. Any test of dowels shall be undertaken on complete assemblies (dowels and sockets).
5. Where the shear resistance of dowels is used in the design the load–deflection ratio of the dowel shall be established on specimens cast into comparable concrete.
6. The testing speed for pull-out and shear tests should not exceed 25 mm/min.
7. The durability of dowels for the intended design life shall be demonstrated to the satisfaction of the Engineer. Where the performance of dowels is only required during tunnel construction it shall be demonstrated that the dowel does not affect the durability of the tunnel to the satisfaction of the Engineer.

8. Dowel assemblies consist of the dowel element and the recess former. Where tensile forces are transmitted, a dowel socket is part of the assembly. The dowel manufacturer shall provide all elements of the product and declare the expected performance.

9. Where the performance of the assembly depends on parameters specific to the design, such as concrete performance characteristics or specific geometry of the installed assembly, the manufacturer shall specify appropriate confirmatory testing for the acceptance of the Engineer. Test results shall be supplied to the satisfaction of the Engineer.

## 217. Caulking and pointing

**217.1. Caulking**

1. Lead for caulking shall comply with BS EN 12588.
2. Lead shall be supplied in rod or strip of widths appropriate to the segment joints, or as lead wool.
3. Cementitious caulking compound cord shall be asbestos-free.

**217.2. Pointing**

1. Mortar for pointing shall be cement:sand (1:3), or otherwise agreed, with water sufficient only to provide a workable consistency that can be built up in suitable layers and packed into the joint. Mortar shall be used within one hour of mixing. Cement shall comply with British Standards as detailed in BS 8500-2 Table 1 as appropriate; sand shall comply with BS EN 12620 and be of a grading commensurate with the work.
2. Additives and proprietary mixes may be used with the Engineer's agreement.

## 218. Timber

### 218.1. General

1. All timber that is used in the Works shall be sourced and procured from a forestry plantation that is subject to the requirements of an internationally recognised Sustainable Forest Management (SFM) Initiative.

2. All timbers used in tunnel construction or underground shall be in accordance with the requirements of BS EN 1995-1-1 and BS EN 338.

3. Details of the proposed use of timber for ground support shall be issued to the Engineer for agreement.

4. All timber that is left in situ in the Works shall be treated. The timber shall be impregnated with preservative fluid in accordance with the requirements of UC4 of BS EN 335.

5. All timber shall be inspected for damage or other strength-reducing factors that may have occurred after the stress grading operation has taken place. Any timber showing such damage shall be indelibly marked as rejected. Reuse of timber shall be permitted but it shall be inspected for damage or excessive deterioration before reuse and, if found unsuitable, rejected.

6. When it is necessary to cut a piece of treated timber for use as ground support, the cut face(s) shall be treated preferably by immersion in the preservative used in the impregnation process. Alternatively, the preservative may be applied liberally by brush.

7. All timber to be used underground shall be treated with a fire-resistant coating agreed with the Engineer, and in accordance with BS EN 1995-1-1 and BS EN 1995-1-2.

8. The use of timber shall meet the requirements of LU Category 1 Standard S1085 Fire Safety Performance of Materials in respect of flammability, smoke emission and toxic fume emission in sub-surface railway (formerly referred to as 'Section 12' areas as defined in The Fire Precautions (Sub-surface Railway Stations) (England) Regulations 2009), shafts and tunnels (both bored tunnel and cut & cover tunnel forms).

9. Timber wedges shall be made of knot-free hardwood.

## 219. Grout for compensation grouting

### 219.1. General

1. Materials supplied in bulk shall be weigh batched, and the quantities per batch recorded. Water and other liquid admixtures used shall be measured to an accuracy of +2% of the target weight, and the records for each mix batch taken and maintained.
2. The final set product must be stable and must not deteriorate over time.

### 219.2. Fluid cementitious grouts

1. Fluid cementitious grouts are defined as those whose range of viscosity and shear resistance is best monitored for consistency by a flow trough, such as the Colcrete standard type.
2. Constituents of fluid cementitious grout shall be ordinary Portland cement to BS EN 197-1 and clean potable water. Admixtures will be permitted to modify viscosity and shear strength and may include such materials as cellulose ethers, PFA or limestone dust.
3. Fluid cementitious grouts shall be prepared in a conventional stirred grout tank or high shear rate mixer and stored until use in a tank with a continuous stirrer.
4. Flow through tests shall be taken for each batch of fluid cementitious grout mixed and results recorded.

### 219.3. Cementitious mortar pastes

1. Cementitious mortar pastes are defined as those with shear strength range appropriate for monitoring by means of the concrete slump tests defined in BS EN 12350-2.
2. Cement mortar paste shall typically comprise ordinary Portland cement to BS EN 197-1 with clean potable water, and combinations of PFA, GGBS, sand, bentonite and cellulose ethers. The design mix shall be set out in the method statement.
3. The mortar paste shall generally have a slump not exceeding 80 mm.
4. Mortar pastes shall be mixed in a concrete mixer and transferred to a remixer for discharge into the pump.
5. Slump tests shall be taken for each batch of mortar mixed and the results recorded. All results shall be made available within 24 h of the tests being completed. All such tests shall be carried out according to British Standard procedures or, if no such standard exists, according to the instructions of the specifier of the test method.

## 220. Maintenance – grouting to control seepage in existing tunnels

**220.1. General**

1. Grouting must only be undertaken using materials approved by the Engineer.
2. All grout materials shall be used in accordance with the manufacturer's guidelines and instructions.
3. All materials used for grouting must have a certificate of conformity.

**220.2. Cementitious grout**

Cementitious grout shall consist of water and cement. The mixture and consistency shall be compatible with the Contractor's equipment and the site conditions.

**220.3. Mortar**

1. Mortar shall be composed of cement and sand mixed in the proportions of 1:3 by volume. An addition of lime up to one quarter of the cement content will be permitted to improve the workability.
2. The quantity of water used shall not exceed that required to produce a mortar of sufficient workability to be placed where required.
3. The ingredients shall be mixed to produce a mortar of medium colour and consistency. Mortar shall be mixed in such quantities as can be used or remixed for use in the Works.
4. No mortar that has become hard or set shall be used or remixed for use in the Works.
5. No frozen materials or materials containing ice shall be used for making mortar.
6. Waterproofing mortar shall be the same as normal bricklaying mortar except that half the water shall be replaced by styrene butadiene rubber latex (SBR).

## 221. Reinjectable grout tubes and grouting

**221.1. General**

1. All reinjectable grout tubes shall have non-return valves along their length to stop ingress into the tube.
2. Reinjectable grout tubes shall not be susceptible to grout holes opening up when they are bent, turned, twisted or otherwise distorted during installation.
3. Where reinjectable grout tubes are part of the permanent works, their ends shall be covered or housed within boxes.
4. The purpose of each grout tube shall be clearly identifiable by appropriate labelling and/or colour coding.

**221.2. Reinjectable grout**

1. The grouting system shall be operated such that it can be flushed after an episode of injection to allow for future injection.
2. The grout used for each operation of the reinjectable grout tube shall be agreed with the Engineer.
3. Each operation of the reinjectable grout tube shall be recorded together with the material used and its associated certificates.

**The British Tunnelling Society**
ISBN 978-0-7277-6643-4
https://doi.org/10.1680/st.66434.099

# 3. Methods

## 301. Excavation for tunnels

### 301.1. General

1. The Contractor shall be responsible for the safety and security of excavations at all times during the execution of the Contract.
2. Mechanised techniques for excavation shall be used wherever practicable to eliminate or reduce health and safety risks.
3. Use of hand-held tools shall comply with the guidance given in *The Management of Hand–Arm Vibration in Tunnelling. Guide to Good Practice* by the British Tunnelling Society.
4. The Contractor shall provide details of their proposed methods for excavation support and spoil removal to the Engineer for agreement. No excavation shall take place until the Engineer's agreement has been obtained. Such agreement shall not relieve the Contractor of any of their obligations under the Contract. The details provided shall include the following

    (*a*) a key excavation plan defining the sequence and direction of all tunnelling that shall be tied to the detailed construction programme. This plan should define personnel/plant access and egress routes for all stages of construction. This should include the route by which a casualty can be evacuated from the Works
    (*b*) safe stop procedures for unplanned stoppages
    (*c*) procedures for managing obstructions (including any existing temporary works) encountered during excavation
    (*d*) procedures for management of excavated material and any waste material within the worksite
    (*e*) procedures for survey control
    (*f*) procedures for water management
    (*g*) procedures for ventilation
    (*h*) measures to protect any operational infrastructure from the works
    (*i*) arrangements for any personnel exclusion zones
    (*j*) arrangements for the storage of materials
    (*k*) arrangements for probing, face logging and management of in-tunnel monitoring.

5. Excavation shall be carried out in a uniform and controlled manner and over-excavation shall be kept to a minimum consistent with the need to maintain the necessary clearance for construction of the Works.
6. The invert of the tunnel shall be protected against damage and deterioration that may be caused by construction traffic. The surface of any sections of invert used as a haul road or

pedestrian walkway shall be maintained in a firm and level condition free from potholes. Any other surfaces that deteriorate or are damaged shall be made good to a standard agreed with the Engineer.

7. Excavation shall be carried out in sections limited to such lengths, depths and widths as may be safely executed having regard to all the circumstances and as appropriate to the ground conditions and the equipment and method of construction being used.

8. In water-bearing strata, the Contractor shall use such methods and take such steps as are necessary to control flows and maintain the stability of the excavation.

9. Where necessary to ensure the safety and security of the Works, excavation shall be continuous by day and night.

10. Planned and enforced stoppages will require the Works to be made safe and inspected by the Contractor at intervals agreed with the Engineer. Remote monitoring methods shall be acceptable.

11. Any voids formed during the excavation process by machine overcut slips, falls of material, overbreak and temporary works shall be filled completely with grout, concrete, sprayed concrete or other approved durable material.

12. Where the Contract specifies limits to surface settlement and/or protection in respect of existing services or structures, the Contractor shall provide calculations demonstrating that the method of excavation will result in compliance with those requirements. Details of the monitoring arrangements that are proposed for the recording of movements and the verification of the degree of any settlement or damage to services or structures shall be in accordance with Clause 329.

13. Where agreed or required by the Engineer, temporary support shall be left in the Works. Untreated timber shall not be left permanently in the Works.

14. The volume of excavated material shall be measured and recorded as the Works proceed. The Contractor shall present to the Engineer at regular intervals specified in the Contract a reconciliation of volumetric advance of tunnel against volume of materials excavated and volume of grout placed.

15. All excavation shall be carried out to a profile as close as possible to the specified excavation line.

16. The Contractor shall be constantly aware of the possibility of slips and ground movement that may be caused by their method or sequence of excavation. The Contractor shall maintain, on site, material and equipment for use in ensuring the stability of the face.

17. The proximity of other tunnels and excavations shall be taken into account when determining the method and sequence of excavation.
18. Where stated on the Drawings or in the Particular Specification, the Contractor shall undertake tunnel excavation, support and grouting to control ground loss to below the specified values.

**301.2. Rock**

1. Where excavation requires the use of blasting, the provisions of Clause 308 shall also apply.
2. On exposure of a fresh rock face, the rock strata in the face walls and roof shall be mapped geologically in accordance with the requirements of the Contract. The Contractor shall provide access for mapping.
3. Any unsound areas of rock shall be reported to the Engineer immediately, together with a record of the Contractor's action. The face shall not be advanced until the area is properly secured.
4. Where an in situ lining is specified, at no point will rock be allowed to intrude within the specified limit as shown on the Drawings.
5. If rock is not self-supporting, any face of exposed ground, where excavation is to be discontinued for whatever reason, shall be supported by timber, rockbolts, sprayed concrete or other means agreed with the Engineer.

**301.3. Soft ground**

1. In ground that is not self-supporting, measures shall be taken to ensure that no undue loss of or softening of the ground occurs at the face, and that there is no run of material from behind supports or lining.
2. In any exposed ground, temporary emergency support shall be available at the face at all times.
3. Ground and groundwater conditions in any exposed face shall be logged.

## 302. Drilling

### 302.1. General

1. Drill rigs shall meet the requirements of BS EN 16228 or BS ISO 18758 as appropriate.
2. Use of hand-held drilling rigs shall be minimised and when used shall comply with the guidance given in *The Management of Hand–Arm Vibration in Tunnelling. Guide to Good Practice* by the British Tunnelling Society.
3. Drilling shall be carried out to provide holes of the diameters, lengths and inclinations required within the deviation limits given on the Specification and Drawings.
4. The Contractor shall keep a written record of all holes drilled, material recovered and water-flows or seepage and provide the Engineer with a copy if required.
5. The contractor shall adopt measures to minimise dust generated, such as wet drilling or dust capture, to prevent release of dust to the tunnel atmosphere.

### 302.2. Blast hole drilling

1. Holes intended to be charged with explosives for blasting shall be drilled in strict accordance with the agreed pattern, in spacing, direction and depth.
2. The accuracy of installation shall be better than ±5° away from the alignment specified and within 50 mm of the specified location on the rock face.
3. Charging of any holes shall not begin until all drilling on that face has been completed.
4. Drilling of relieving holes following a misfire shall only be done under the direct supervision of an authorised shot-firer.

### 302.3. Probe hole drilling

Probe hole drilling shall be in accordance with Clause 310.

## 303. Temporary works

### 303.1. General

1. Temporary works are as defined in Clause 102. They shall be designed, erected and managed in accordance with the requirements of this specification and the relevant standards.
2. Geotechnical temporary works shall be undertaken in accordance with BS 6164:2019 Clause 6.4.
3. Structural temporary works shall be undertaken in accordance with BS 5975:2019.

### 303.2. Design of temporary works

1. The design and execution of temporary works in respect of their structural safety and robustness shall be undertaken in exactly the same way as if they were permanent works.
2. Design check procedures shall apply equally to both permanent and temporary works. The requirements of clause 6.4 of BS 6164:2019 shall be adhered to that in turn require adherence to requirements set out in BS 5975:2019.
3. Any temporary or permanent works that at any time support the ground shall be subject to a check of at least category 2.
4. Temporary works design shall take account of guidance in *The use of European Standards for Temporary Works design* published by the Temporary Works forum. British Designated standards shall be used where they exist in preference to the equivalent European Standard.
5. Materials used for temporary works shall comply with the requirements of BS 5975 and the materials specifications defined elsewhere in this Specification.
6. Calculations and Drawings shall be submitted to the Engineer for their agreement, as part of a comprehensive method statement, covering the temporary works.

### 303.3. Supervision

1. A group shall be set up comprising the respective 'designated individuals' (a role defined in BS 5975) from all parties to the tunnelling project including the Employer. This group shall be responsible for

   (a) overseeing the design and execution of temporary works and ensuring it is in accordance with BS 5975:2019, Table 22
   (b) collectively agreeing the scope of the checks required for temporary works and the category of check required. In the event of disagreement, the final decision shall be made by the designated individual representing the principal Designer
   (c) establishing and maintaining a register of temporary works for the project in accordance with clause 13 of BS 5975:2019

    (*d*) ensuring the arrangements for the supervision and inspection of temporary works during their execution is in accordance with the requirements of BS 5975

    (*e*) timely reporting to the Employer of concerns where the requirements of this Clause are not being met.

2. The execution of temporary works including the appointment of competent persons to manage their design and execution shall conform to BS 5975.

3. The Principal Contractor shall appoint a Temporary Works Coordinator (TWC) who shall have overall responsibility for the safe design and construction of all temporary works on the project. In addition, the Principal Contractor shall appoint a sufficient number of Temporary Works Supervisors to supervise the construction of temporary works on a day-to-day basis including shift work.

4. The scope of the term *temporary works* includes but is not limited to the items listed in clause 6.4.4 of BS 6164:2019.

5. Temporary works shall not be loaded until a 'permit to load' has been issued by the TWC.

6. Full facilities shall be provided for the Engineer to inspect work in progress.

## 304. Spiling, dowelling and rockbolting

### 304.1. Spiling

1. Where spiling is employed to provide support for advancing the excavation, spiles shall be driven into the ground or placed in pre-drilled and grouted holes as specified on the Drawings or by the Engineer.
2. The embedded length shall be not less than that specified by the Designer.
3. The accuracy of installation shall be better than ±5° away from the alignment specified.
4. The number, location, overlap and angle of spiles shall be commensurate with the ground conditions and methodology specified on the Contract Drawings or by the Engineer.
5. Details of spiling shall be agreed with the Engineer in accordance with Clause 301.1.4.
6. Spiles shall be installed such that a tensile bond is formed between the spile, the ground ahead of the proposed excavation and the sprayed concrete lining. Typically, this may involve grouting the spile into the hole for pre-drilled and self-drilling spiles or hammering the spile in.
7. Care shall be taken during installation of spiles to ensure minimum disturbance of the ground due to the installation process.
8. Probing shall be carried out in conjunction with spiling in order that fully embedded spiles are installed prior to the required location.
9. The provisions of Clause 304.1 can be applied to canopy tubes where confirmed by the Designer.

### 304.2. Rock dowelling

1. Rock dowels shall be inserted into pre-drilled holes. The number, location, length and angle of inclination shall be commensurate with ground conditions. The diameter of holes shall be designed to allow for installation of dowels and any grout and breather tubes to ensure effective transfer of shear stress from rock to the dowel.
2. Holes for the installation of dowels shall be drilled straight and with an accuracy of ±10°. On completion of each hole, and prior to the installation of each dowel, holes shall be cleaned to remove debris. Where water flush is employed, the amount of water shall be kept to an absolute minimum.
3. A regular surface shall be provided to seat the face plate by trimming rock surfaces or forming pads of quick-setting mortar. Where mortar pads are required, they shall be of adequate thickness and extend beyond the face plate by 25 mm all round at that thickness before being chamfered at 45°. Care shall be taken to ensure that the mortar does not interfere with the installed dowel.
4. Cementitious grouting material shall be injected starting from the furthest point of the drilled hole so that the dowel is

completely encased in grout. The grouting procedure is dependent on orientation to the horizontal. A breather tube or equivalent shall be used. Grout shall not be used after a period equivalent to its initial setting time. Where cement grout is used, a set of six cubes of cement grout shall be taken when each series of rock dowels is in progress. Sampling, preparation, curing and testing shall be in accordance with BS EN 196. Half the cubes shall be tested at 1 day and the remainder at 28 days. The average compressive strength determined from any group of cubes shall exceed the specified characteristic strength by

$1 \text{ N/mm}^2$ for cement grout tested at 1 day
$3 \text{ N/mm}^2$ for cement grout tested at 28 days.

5. Resin grouting capsules shall be installed as directed by the manufacturer.
6. The proposed pattern for dowels and full details of the dowels and installation thereof shall be agreed with the Engineer. Where required, dowels shall be installed in the pattern and locations and manner shown on the Drawings.

## 304.3. Rockbolting

1. Where rockbolts are employed, full details of the bolts and installation shall be agreed with the Engineer. The pattern and type of rockbolt and the length and diameter of bolts shall be commensurate with the rock characteristics. Where required, rockbolts shall be installed in the pattern and locations and manner shown on the Drawings.
2. Where required by the Contract, the Contractor shall carefully inspect the excavation as it progresses, and if conditions arise that require a change in the excavation and/or support systems, then revised proposals shall be agreed with the Engineer.
3. Rockbolts shall be provided with an even and secure bearing for their face plates. Holes shall be drilled to produce straight holes of the required length, and with an accuracy of $\pm 10°$, and cleaned out by flushing with compressed air or with clean water to remove debris prior to fixing the rockbolt. The amount of water-flushing shall be kept to an absolute minimum.
4. Grouting of the rockbolts shall take account of the following.

    (*a*) Inclined rockbolts.
    
    (*i*). Flowing grouts may be placed by an injection pipe at the mouth of the borehole against packing material to seal the mouth and require a breather tube to prevent airlocks within the hole.
    (*ii*). Thixotropic grouts are typically placed by tremie pipe to the distal end of the borehole, with the grout flowing back down the annulus of the borehole.

(b) Declined rockbolts.
  (i). A single tremie pipe to the base of the borehole for solid bars or a self-drilling hollow bar installed using simultaneous drill and grout.

(c) Horizontal rockbolts.
  (i). Bolts should be ±10° to ensure that grout can be trapped in the borehole either by gravity or through the use of packers on inclined bores.

5. Rockbolts shall be fixed as soon as possible after excavation. A selection of rockbolts shall be tested. Where excavation is progressed by blasting, any rockbolts within three metres of the face shall be retested. Any rockbolt that fails a test shall be adjusted to restore its nominated load or replaced as necessary.
6. The Contractor shall carry out in situ tests as agreed with the Engineer. Testing is to be generally in accordance with the procedures given in ISRM Document 2, Part 1 *Suggested methods of rockbolt testing*.
7. Unless otherwise provided for in the Contract, at least 5% of the first 100 bolts installed shall be subject to test. Thereafter 2.5% of subsequent bolts shall be tested. The remedial actions following any bolt test failures shall be agreed with the Engineer.

## 304.4. Load testing

1. Before commencing rockbolting or dowelling a test programme shall be undertaken to establish the capacity of the rockbolts/dowels to provide the support required.
2. Suitability load testing and acceptance load testing for rockbolts and dowels shall be undertaken in accordance with an agreed quality control programme as set out in the Drawings or the Particular Specification.

## 304.5. Records

For each dowel or rockbolt installed, the following information shall be recorded

(a) reference number
(b) type of installation
(c) name of person taking record
(d) date of drilling
(e) length and orientation of drill hole
(f) consistency, colour, structure and type of rock or material
(g) rates of penetration and water flow if any
(h) date of dowel or bolt installation
(i) length of installation
(j) diameter of hole
(k) diameter of bolt/dowel
(l) type and volume of grout used/injected

(*m*) water/cement ratio of the grout

(*n*) type and volume of additives (if any)

(*o*) pressure of injection (if flowing grout is used)

(*p*) details of tests, where carried out.

## 305. Sprayed concrete

### 305.1. General

1. Sprayed concrete shall be applied by the wet process. Exceptionally, the dry process may be used where necessary for the safety of the works. All aspects of the application of sprayed concrete shall be subject to the agreement of the Engineer. Particular emphasis shall be placed on the control of dust emissions to the tunnel atmosphere and the guidance in BS 6164:2019 clause 16 shall be adhered to.

2. The Contractor shall develop a sprayed concrete mix and a plan for its production and application including the control of emissions. Constituent materials shall comply with the requirements of Clause 210 of this Specification.

3. The sprayed concrete shall be developed in two stages

    (*a*) the production of a suitable unretarded and unaccelerated base concrete

    (*b*) the production of sprayed concrete from the base concrete.

4. The sprayed concrete mix design shall, unless otherwise stated, comply with the characteristic strengths specified by the Designer for early-age and long-term loading.

5. For conformity control of sprayed concrete, three inspection categories have been specified (see BS EN 14487-1:2022 Clause 7). Temporary sprayed concrete linings (those used for temporary works) fall into Inspection Category 2, and permanent sprayed concrete linings (those used for permanent works) fall into Inspection Category 3 (see BS EN 14487-1:2022 Annex A, Table A.3). Refer to Clause 102 for definition of permanent and temporary works.

6. The Designer shall be represented on site at all times during tunnel construction to verify that the materials and workmanship are consistent with the design, and to ensure that ground and groundwater conditions are in accordance with design assumptions. The Contractor shall establish a procedure to respond effectively to changes in ground and groundwater conditions from the design assumptions (see also Clause 329 of this Specification).

7. The Contractor shall establish and maintain the instrumentation and monitoring required by the design. The Contractor shall establish a procedure (the 'Instrumentation and Monitoring Plan') that will enable prompt and regular review and effective response to the results from the instrumentation and monitoring. The Instrumentation and Monitoring Plan shall be subject to the Designer's agreement. The Designer shall be included in the monitoring review procedure (see also Clause 329 of this Specification).

8. The Contractor shall submit the concrete mix design certificates to the Engineer for acceptance, demonstrating compliance of all constituents with relevant standards and this Specification prior to their use in the Works.

9. Where a concrete mix design has been used for similar applications on previous projects, the Contractor may submit historic testing data from these projects to demonstrate to the Engineer the compliance of constituents with the design, the relevant standards and this Specification.

10. The frequency of tests undertaken as part of the conformity testing may be adjusted subject to the agreement of the Engineer where a concrete mix has demonstrated a high level of compliance with all testing requirements. Supporting evidence may include historic test data from previous projects if the concrete mix has been used across multiple projects for similar applications. Conversely should test results demonstrate non-compliance with design requirements, or a trend towards non-compliance, then a higher frequency of testing shall be required.

## 305.2. Proficiency

1. Sprayers shall hold relevant certificates of competence issued by the Contractor or written evidence of previous satisfactory work indicating compliance with EFNARC Nozzleman Certification Scheme or similar National Standards to the approval of the Engineer.

2. Sprayed concrete linings in the Works shall only be applied by Sprayers who have successfully demonstrated their competence to the agreement of the Engineer. At least one vertical and one overhead acceptance panel shall be required per Sprayer. Tests for proficiency shall use the equipment selected for use in the Works where practicable.

3. Where bar reinforcement is used, each Sprayer shall also demonstrate acceptable proficiency in the application of sprayed concrete to reinforcement in trial areas before doing so in the Works.

4. In addition to the competence demonstration of individual sprayers, each crew shall demonstrate acceptable proficiency in the application of sprayed concrete to trial areas, including competence in batching and pumping, before being employed on the Works to the agreement of the Engineer.

5. Test panels shall be sprayed using the same Sprayers, batchers, pump operators, assistants, equipment and approved concrete mixes (including specific constituents) that will be used for the Works. If the sprayed concrete for the Works is to be supplied from a pre-batched silo system, then the test panels must be supplied from a similar pre-batched silo system using equivalent screw arrangement and silo volume.

6. The Contractor shall submit an organisation chart and résumés of key personnel undertaking the ground works to the Engineer for approval. The key personnel shall include the Underground Manager responsible for the Works, Shift Engineers, Tunnel Shift Foremen and Sprayed Concrete Sprayers. The organisation chart shall show how the Designer integrates into the management structure. Changes to these personnel shall also be approved by the Engineer.

7. The Contractor shall establish a procedure for continuous assessment of Sprayers and other operatives including review of test cores and samples. Where repeated failures in production are attributed to individuals, the Contractor shall reassess their competency and provide retraining as required to ensure that the specified quality can be achieved.

## 305.3. Batching and mixing

1. Batching and mixing shall be carried out by equipment capable of properly mixing materials in sufficient quantity to maintain the continuous application of sprayed concrete and to the accuracy defined in BS EN 14487-2.

2. All measuring equipment shall be maintained in a clean serviceable condition and shall be zeroed daily and calibrated regularly.

3. Refer to Clause 203.3 for requirements for use of fibres.

4. Full and complete records shall be kept available for inspection at the request of the Engineer of all constituent materials for the sprayed concrete.

5. Hand dosing of any of the mix constituents shall not be permitted, except for trial mixes where automatic dosing is not practicable.

6. Measures shall be taken to prevent any adverse impacts on the quality of the sprayed concrete, its constituents and admixtures due to cold or hot weather.

## 305.4. Application

1. Details of all equipment to be used shall be made available to the Engineer prior to commencement of site trials. The sprayed concrete nozzle and ancillary equipment shall be of an adequate capacity for the volumes to be applied.

2. The equipment selected by the Contractor and approved by the Engineer will be capable of maintaining the ratio of concrete and accelerator as selected from the trials and approved by the Engineer. The actual ratio of accelerator to concrete selected shall be identified at the nozzle and take into account the filling efficiency of the equipment and the efficiency of the accelerator dosage equipment to overcome the air and concrete pressure at the nozzle while spraying at typical outputs and air flows.

3. A complete stand-by sprayed concrete system of plant and ancillary equipment in full working order shall be available for the duration of the tunnel excavation. The system shall be agreed by the Engineer and as a minimum shall be capable of implementing all safe stop procedures.

4. The sprayed concrete system shall supply air, water and other constituents uncontaminated by material deleterious to sprayed concrete. In particular, care shall be taken to exclude oil from the air supply. Mineral oil or other formwork release oils used to protect spraying equipment shall not be used in pump hoppers, re-mixer drums or concrete skips where their presence can become incorporated into the concrete mix.

5. Equipment shall be thoroughly cleaned at least once per shift. The spray nozzle shall be checked for wear and where necessary replaced.

6. Transport pipes consisting of hoses and pipes shall be designed to convey the concrete efficiently and without leakage or blockage. The transport pipes shall have uniform diameter appropriate to the mix characteristics determined by site trials and be free from any dents or kinks between the sprayed concrete machine and the nozzle.

7. The working area for sprayed concreting shall be well illuminated and ventilated. Dust pollution shall be minimised by choice of appropriate equipment and by means of additional ventilation, water sprays, and by maintaining equipment in good order and shall comply with the requirements of the local standards and regulations. Dust emissions shall be controlled within local statutory limits. Protective clothing and respiratory protective equipment/dust masks shall be provided for and used by all persons present during spraying.

8. The Contractor shall produce and enforce an exclusion zone policy to protect the workforce during sprayed concrete spraying.

9. Spraying shall only be conducted within a strict exclusion zone where only trained and competent persons authorised by the exclusion zone policy are able to enter.

10. The equipment shall allow for air and water in any combination to be available for preparation of surfaces and/or cleaning of finished work.

11. Dosing of admixtures by hand shall not be permitted.

12. A robotic spray machine operated remotely at a safe distance from the face shall be used unless it can be demonstrated to the Engineer that the use of such equipment is impractical.

13. The Contractor shall allow the Engineer access to the sprayed concrete Works at all times and shall allow the Engineer access to inspect the excavated ground surface prior to spraying if requested, within the rules of the exclusion zone policy.

14. The method of spraying should ensure a high-quality product (see BS EN 14487-2 Clauses 5.1.1, 5.1.2, 5.2.2 and 9.1). In particular, the spray nozzle shall be kept as perpendicular as possible to the surface and care shall be taken to achieve a regular properly compacted coating of the correct thickness.

15. Spraying of the Works shall commence when the air accelerator and concrete materials are all flowing in unison through the nozzle. In no circumstances shall air and accelerator be directed into the Works without the addition of concrete at the nozzle. The sprayed concrete shall emerge from the nozzle in a steady uninterrupted flow. Should the flow become intermittent for any cause, the Sprayer shall direct it away from the Works until it again becomes constant.

16. A means of profile control shall be used that enables spraying without the need for personnel to access under unsupported ground or fresh sprayed concrete. The thickness and position of the sprayed concrete shall be defined by screed boards, lattice arches, guide wires, depth pins, lasers or other means.

17. For vertical and near-vertical surfaces application shall commence at the bottom and the leading edge of the work shall be maintained at a slope. Downward spraying shall be avoided where possible. The nozzle may be inclined sufficiently to ensure reinforcement is properly embedded.

18. Prior to continuation of spraying from a joint or leading-edge position or in any other circumstances where sprayed concrete has hardened beyond its initial set, loose material shall be removed by jetting with a compressed air lance. Any laitance that has been allowed to take final set shall be removed and cleaned by jetting with air and water.

19. All rebound material shall be removed from the working area and shall not be used in the Works.

20. All joints in the sprayed concrete lining shall be as specified in the design.

21. The surface to receive sprayed concrete shall be damp but shall not exhibit free water.

22. Where groundwater flow could interfere with the application of sprayed concrete or cause reduction in the quality of sprayed concrete the Contractor shall take all action necessary to control groundwater. Such action shall include the channelling of water by means of pipes and chases.
On completion of spraying, the pipes and chases shall be plugged and grouted to prevent further water ingress.

23. Sprayed concrete shall be left in its natural finish without further working except as required to trim excess thickness, where the sprayed concrete shall be allowed to stiffen sufficiently before being trimmed with an approved cutting screed.

24. All freshly sprayed concrete shall be adequately protected against the weather and other harmful effects. The temperature of the mix before placing shall not be below 10°C and shall not exceed 35°C unless special provisions are made that have been agreed with the Engineer. Spraying shall not be undertaken when ambient temperature is below 5°C unless special measures can be taken to provide protection against frost until the sprayed concrete has developed a compressive strength of at least 5 MPa.

25. The pot life of the concrete shall be determined by preconstruction trials. Any unused material after this time shall be discarded.

26. The Contractor shall verify the thickness of any sprayed concrete layer at any location requested by the Engineer. Any holes in the sprayed concrete shall be backfilled with non-shrink mortar with characteristic strength equal to or higher than the sprayed concrete, unless otherwise specified by the Designer. The integrity of the waterproofing system shall be maintained at all times.

27. Surfaces that are not to receive sprayed concrete shall be protected.

28. Special care shall be taken while spraying concrete around lattice, mesh and bar reinforcement and embedded structural steel sections to ensure full and complete placement of concrete without voids caused by shadowing.

29. Where an initial layer is specified by the Designer it shall be immediately applied in one pass on all exposed ground after excavation is complete and the profile has been checked.

30. For full face excavation and spraying, the invert shall be completed to full thickness and allowed to set adequately to remove rebound before commencing spraying of arch and crown.

31. For invert lining sprayed concrete application, a compressed air lance shall be used during spraying to remove rebound from the spraying area.

32. Sprayed concrete shall first be placed in corners, recesses and other areas where rebound or overspray cannot escape easily. Place sprayed concrete with nozzle positioned approximately perpendicular to the receiving surface. In corners, direct nozzle at an approximately 45° angle or bisect the corner angle.

33. Sprayed concrete shall be applied so sags or sloughing does not occur. Where movement of sprayed concrete has occurred adjacent to a slough-off, the sprayed concrete in question shall be removed as directed by the Engineer.

**34.** The minimum strength that sprayed concrete must achieve before any activities are carried out below a sprayed area shall be identified in the design. This will input to personnel exclusion zones and safe working areas that shall be defined for all sprayed concrete activities. This may be determined by tests to BS EN 14488-2 or by using temperature measurements coupled with a maturity method calibrated to the mix being used.

**35.** Sprayed concrete shall be applied in such a way as to avoid delamination or fall-outs. The method of application shall be agreed with the Engineer.

## 305.5. Curing

**1.** Sprayed concrete shall be cured in accordance with BS EN 14487-2 Clause 9.3 unless otherwise approved by the Engineer.

**2.** Proprietary curing compounds or methods may be used only with the agreement of the Engineer. The use of surface applied curing membranes shall be selected so that they maintain or improve the bond to subsequent concrete layers. The use of curing membranes that negatively influence bond shall be reserved only for final concrete surface applications.

## 305.6. Defects and repairs

**1.** Before a subsequent layer of sprayed concrete is placed, the preceding layer shall be checked for defects.

**2.** Areas of work shall be properly compacted and bonded and free from honeycombing, laminations, dry or sandy patches, voids, sagged or slumped material, rebound, excessive cracking and overspray.

**3.** Where defects occur, the Contractor shall agree with the Designer and the Engineer proposals for the removal of the defective material and replacement by material without defect. Where a defect is required to be rectified, the area to be replaced shall in any event be not less than 300 mm × 300 mm.

## 305.7. Reinforcement

**1.** Reinforcement shall be securely fixed to avoid movement or vibration during spraying. BS EN 14487-2 Clause 6 shall be adhered to.

**2.** Where two layers of reinforcement are to be incorporated in the work, the rear layer shall be sprayed prior to fixing the front layer.

**3.** Lattice girders shall comply with Clause 208.3 and Clause 306.1 of this Specification.

**4.** The ability to successfully encapsulate the maximum design reinforcement shall be proven by preconstruction trials.

**305.8. Tests for trial mixes**

1. The Contractor shall propose to and agree with the Engineer trial mixes for the Works at least 56 days before their commencement. Tunnelling shall not be permitted to start until the preconstruction tests have been approved by the Engineer.
2. Compressive strength at 7 days and 28 days and 28-day ultimate flexural strength testing shall be undertaken for a non-accelerated mix, for all mixes trialled.
3. For mixes and spraying equipment that have been validated on previous projects and where compliance with all project specifications can be clearly documented to the Engineer, the extent of trials can be limited to trial panel spraying to demonstrate mix design conformity.

**305.9. Test panels and acceptance tests**

1. The performance requirements shall be set by the Designer.
2. The Contractor shall undertake preconstruction tests by means of test panels to demonstrate the performance requirements can be met and as acceptance tests to prove the competency of Sprayers. Each mix will require individual trials.
3. The Contractor shall provide a report detailing the proposed trial arrangements in advance of the trials to the Engineer for their acceptance. The trials shall be undertaken sufficiently in advance of commencement of sprayed concrete works to allow acceptance, see Clause 305.8.1.
4. For each type of sprayed concrete to be used, a trial mix shall be designed by the Contractor and prepared with the constituent materials in the proportion proposed. Target workability (consistence) values shall be determined for the wet mix process. Sampling and testing procedures shall be in accordance with BS EN 12350. A clean, dry mixer shall be used and the first batch discarded.
5. The trials shall replicate the tunnel environment in which spraying will be undertaken. Dust levels shall be monitored in real time in accordance with BS 6164:2019 clause 16 at the location of the Sprayer and at points 5 and 10 m from the Sprayer. Dust levels shall not exceed the limits in BS 6164:2019 clause 16.3 at the 10 m point. Spraying shall last at least 20 min for a trial to be considered valid.
6. The Contractor shall use the results of the trials to inform the design of the ventilation regime required to give adequate control of dust emissions in the tunnel.
7. The equipment proposed for the application of concrete in the Works shall be used for the trial. The trial will establish whether the selected equipment is capable of efficiently mixing concrete, accelerator and air at the nozzle, and be capable of positioning the nozzle at a suitable distance and orientation to the surface geometry of the structure to which the concrete is to be applied.

8. During the trials the Contractor will establish the volume of air required to give adequate compaction of the material using the nozzle and conveyance lines selected for the Works. If the delivery equipment or nozzles are to be changed during the course of the Works, the volume of air required will need to be verified again. The equipment will be maintained adequately to ensure that the required volume of air can be maintained while spraying. Air pressure can only be used as a control if the air delivery system is not altered from the original verification trial. No additional taps or restrictions will be permitted to be added into the system without repeating the verification trials.

9. If a particular quality of finish is required other than as sprayed, the trials will evaluate the methods and tools to be used to achieve the required finish and the Engineer will approve the method and quality of finish achieved.

10. From the trial mix an experienced Sprayer shall prepare sufficient test panels as per BS EN 14488-1 clauses 4.3 and 5.4. The sprayed concrete in the panels shall adhere well to the backform, be properly compacted and exhibit no sagging.

11. If a sprayed secondary lining is to be applied in combination with a waterproofing membrane, a mock-up shall be prepared to demonstrate proper installation of the sprayed concrete secondary lining on to the waterproofing membrane to the satisfaction of the Engineer.

12. The panels shall not be moved for 18 h after spraying and shall be stored without disturbance at a temperature of 15–25°C and covered by polythene sheet until the time of coring.

13. Cores for strength tests shall be obtained from the panels at 1 day in accordance with BS EN 14487-1. The cores shall be visually inspected and the dimensions and comments regarding the quality of the cores shall be recorded as required in BS EN 12504-1. Cores shall not be taken closer than 125 mm from any edge of the panel. The cores shall be free from lamination. The cores for strength tests shall be stored in water.

14. The testing requirements shall be compressive strength in spray direction after 1 day and 28 days on a set of five cores each as per BS EN 14487-1. The correction for in situ compressive strength is defined in BS EN 13791. For each test at least one spare specimen shall be provided. No sets of cores to be tested at any given age shall come from the same panel. Cores to be tested at different ages – that is, 1 day and 28 days – may come from the same panel.

15. The Contractor shall carry out other tests and trials during the period of the preconstruction trials as specified in BS EN 14487-1 Clause 7.3 to confirm that the proposed mixes and

methods meet the minimum performance requirements, unless otherwise agreed with the Engineer.

16. If required by the Engineer, the trial shall include the construction of the proposed joints including layer joints and advance joints. Samples taken from across the joints will then be taken and the results reported.

17. The site trials, including dust emissions trials, shall be repeated if the source or quality of any of the materials, mix proportions or placing equipment is to be changed during the course of the Works.

18. Should any mix fail to produce satisfactory sprayed concrete, the Contractor shall repeat the construction of test panels and test either the same mix, plant and labour or make such adjustments as they consider necessary.

19. Should any dust trial fail to demonstrate adequate control of dust emissions, the Contractor shall make such adjustments to further reduce emissions as they consider necessary and repeat the trial until successful.

20. Flexural beam testing shall be performed at an age of 28 days and as otherwise agreed with the Designer. Tests shall be in accordance with BS EN 14651. A minimum of three flexural beam test results shall be obtained for each tested fibre trial mix. The characteristic residual flexural tensile strength shall be established in line with BS EN 1990 D7.2

21. Each acceptance test shall include as a minimum one test panel as vertical and one overhead (minimum 45° above horizontal).

22. The trials shall be repeated if the mix constituents are changed from the approved mix design.

23. Full test reports detailing each panel and associated tests shall be made available to the Designer and Engineer for approval.

**305.10. Production tests**

1. Tests shall be carried out on a routine basis as agreed with the Engineer on cores taken from sprayed concrete placed in the Works. Where a reliable correlation between concrete strength and other parameters has been established, production control based on compressive strength results is acceptable subject to the agreement of the Designer and the Engineer. The location of the cores shall be determined to give even distribution around the entire lining – that is, samples from crown, axis and invert. The integrity of the waterproofing system shall be maintained at all times.

2. Where the nominal required sprayed concrete thickness is less than 100 mm, the cores for the compressive strength testing shall be taken from areas where the actual thickness is greater than 100 mm. Alternatively, additional sprayed concrete shall be applied in selected areas agreed by the Engineer for subsequent coring of test specimens.

3. Compressive strength tests shall be carried out on prepared test cores in accordance with BS EN 14487-1 and BS EN 14488-2. The time of coring shall be as close as possible to 24 h after placing. Cores required for 28-day strength tests shall be obtained at the same time as those for 1-day tests and stored in the laboratory in accordance with BS EN 12504-1 and BS EN 12390-2.
4. The frequency of coring shall be in accordance with BS EN 14487-1. The minimum sampling frequencies are valid for production volumes or areas as indicated in Table 10. For volumes or areas smaller than those in Table 10, at least one test sample shall be taken.

Table 10 Control of sprayed concrete properties (see BS EN 14487-1)

| Method No. | Type of test | Inspection/test according to | Minimum sampling frequency Strengthening of ground | | |
| --- | --- | --- | --- | --- | --- |
| | | | Category 1 | Category 2 | Category 3 |
| **Control of fresh concrete** | | | | | |
| 1 | Water/cement ratio of fresh concrete when using wet mix method | Calculation or test method | | | Daily |
| 2 | Accelerator | Record of quantity added | | | Daily |
| 3 | Fibre content in the fresh concrete | BS EN 14488-7 | Min 1 | $1/200\,m^3$ or $1/1000\,m^2$ | $1/100\,m^3$ or $1/500\,m^2$ |
| **Control of hardened concrete** | | | | | |
| 4 | Strength test of young sprayed concrete | BS EN 14488-2 | $1/5000\,m^2$ or 1/2 months | $1/2500\,m^2$ or 1/month | $1/250\,m^2$ or 2/months |
| 5 | Compressive strength | BS EN 12504-1 | $1/1000\,m^3$ or $1/5000\,m^2$ | $1/500\,m^3$ or $1/2500\,m^2$ | $1/250\,m^3$ or $1/1250\,m^2$ |
| 6 | Density of hardened concrete | BS EN 12390-7 | When testing compressive strength | | |
| 7 | Resistance to water penetration | BS EN 12390-8 | 1 | 1/6 months | 1/month |
| 8 | Freeze–thaw resistance | As no European Standard on this issue is available at the publication of this document, National Standards apply | 1 | 1/6 months | 1/month |
| 9 | Bond strength | BS EN 14488-4 | | $1/2500\,m^2$ | $1/1250\,m^2$ |

Table 10 (*Continued*)

| Method No. | Type of test | Inspection/test according to | Minimum sampling frequency | | |
|---|---|---|---|---|---|
| | | | Strengthening of ground | | |
| | | | Category 1 | Category 2 | Category 3 |
| **Control of fibre-reinforced sprayed concrete** | | | | | |
| 10 | Fibre content of hardened concrete (this test is alternative to method No. 3 when it is not practicable to determine the fibre content from the fresh sprayed concrete) | BS EN 14488-7 | When testing residual strength or energy absorption capacity | | |
| 11 | Residual strength or energy absorption capacity | BS EN 14651 or BS EN 14488-5 | $1/2000 \, m^3$ or $1/10 \, 000 \, m^2$ | $1/400 \, m^3$ or $1/2000 \, m^2$ | $1/100 \, m^3$ or $1/500 \, m^2$ |
| 12 | Ultimate flexural strength | BS EN 14651 | When testing residual strength | | |

5. The strength of sprayed concrete measured by cores taken from the Works shall be acceptable if the compressive strength results comply with the requirements stated in Clause 210.10 of this Specification or alternative requirements specified by the Designer.

6. Mechanical rebound hammers shall not be used to obtain indirect compressive strength of sprayed concrete.

7. If the results of any production test do not comply with the evaluation criteria, the results and tests procedures shall first be checked and confirmed. The Engineer may require that additional tests be carried out by the Contractor to determine the extent of the non-compliance and/or new mix proportions or methods determined to avoid further failures.

8. Where sprayed concrete does not comply with the required strength, the Contractor shall execute remedial work, which may involve additional sprayed concrete or replacement in sections where it is safe to do so. The Contractor shall take into account any limits placed on the tunnel profile dimensions resulting from the Specification. The Contractor shall submit to the Engineer for agreement a method statement, specification and calculations for remedial work.

9. The Engineer shall, in the event of repeated failure in quality control, require the Contractor to adjust the mix to achieve the required strength.

10. The Contractor shall keep a record in a form to be agreed with the Engineer of all tests on sprayed concrete, which shall be kept on site identifying the tests with the section of work to which they relate.

11. Site-specific calibration is required for the strength tests of young sprayed concrete in BS EN 14488-2.

12. For permanent linings, test panels sprayed in the tunnel can be used instead of taking samples from the lining. These shall be sprayed immediately after spraying the layer to which they apply. At least one spare test specimen shall be provided from the test panel. In addition, at least one spare test box shall always be available near the face being sprayed.

13. All core holes in the lining should be backfilled with non-shrink mortar of an equivalent strength unless otherwise specified.

14. Where cores are taken from test panels, at least one core per three panels shall be taken along the panel, parallel to the front and back of the panel. These cores shall be acceptable if the compressive strength results comply with the requirements stated in Clause 210.10 of this Specification.

## 305.11. Acceptance

1. Deficiencies observed during the sprayed concrete application process such as, but not limited to, the following constitute a cause for sprayed concrete rejection

   (*a*) failure to properly control and remove build-up of overspray and rebound

   (*b*) incomplete encasement of reinforcing steel and spiles

   (*c*) incorporation of sand lenses, voids, delamination, sagging or sloughing

   (*d*) failure to apply sprayed concrete to the required line, gradient and tolerance.

2. Hardened sprayed concrete will be visually examined by the Engineer for any evidence of excessive drying shrinkage cracking, tears, feather edging, sloughs or other deficiencies. The Contractor shall undertake sounding to check for delaminations.

3. If the results of the compliance tests from sprayed concrete test panels or the assessment of the fresh or hardened sprayed concrete indicate non-conformance of the sprayed concrete, the Contractor shall implement a programme of evaluation of the in-place sprayed concrete for the Engineer's approval. Such evaluation shall include but not be limited to the following

   (*a*) extraction of cores from the linings at locations selected by the Engineer and testing of such cores for compliance

   (*b*) checking for delaminations using sounding, hammer tapping or other appropriate non-destructive testing procedures

(c) bond pull-off testing

(d) diamond saw cutting or coring to check adequacy of encasement of reinforcing steel

(e) checking of as-built lining thicknesses.

The Contractor shall provide a report to the Engineer defining the findings and proposals for mitigation of the defective works and amendments to the working methods/materials to avoid recurrence. This process shall work in parallel with non-conformance reporting.

4. Rejected sprayed concrete shall be replaced by compliant sprayed concrete to the satisfaction of the Engineer or the lining strengthened as approved by the Engineer. The remedial works shall be reviewed at the Daily Review Meeting (DRM) and be subject to a separate Required Excavation and Support Sheet (RESS).

**305.12. Delivery, storage and handling**

1. Materials shall be delivered, stored and handled to prevent contamination, segregation, corrosion or damage.

2. Admixtures shall be stored in accordance with the manufacturer's recommendations.

3. Liquid admixtures shall be stored to prevent evaporation and freezing. Admixtures shall be stored in clearly marked and labelled containers at all times (including admixture name, type, storage requirements, use-before date, instructions for use, safety precautions and manufacturer's recommended dosage range).

## 306. Ground support with arches, ribs and lattice girders

**306.1. Steel arches/ribs and lattice girders**

1. The Contractor shall provide method statements for the erection of arches to the Engineer for agreement. Arch materials shall conform to Clause 208.

2. Steel arch ribs and full circle ribs shall be firmly fixed in their final positions against the excavation. Arch bases shall be provided with integral base plates of size to suit the bearing capacity of the ground and shall bear on rock or concrete of adequate strength. Arches and ribs shall be sufficiently clear of the excavation and the final internal profile of the structure to accommodate any required concrete cover.

3. The number of joints in the arch shall be varied to suit the Contractor's method of working, subject to the Engineer's agreement. Steel tie bars and struts between arches shall be installed where shown on the Drawings.

4. An adequately thick sprayed concrete layer must be in place before the installation of the lattice girders. Under no circumstance shall the installation of lattice girders take place from under unsupported ground.

5. Lattice girder segments shall be secured by use of temporary wood blocking, steel wedges, concrete spacers, mortar sacks and/or other appropriate means to maintain position during application of sprayed concrete. The means of support shall be subject to the approval of the Engineer. All wood blocking shall be removed during the next phase of excavation and any void infilled with sprayed concrete to ensure continuity of the primary lining.

6. Lattice girder segments shall have butt plates and the method of installation shall ensure tight connection of all elements.

7. Immediately prior to concreting, casting or spraying, the arches, ties and struts shall be rendered clean and free from deleterious matter.

8. Lattice girders shall be firmly fixed in their final position against the excavation prior to application of sprayed concrete. Lattice girders shall be sufficiently clear of the excavation and final internal profile of the structure to accommodate the required sprayed concrete cover.

9. Lattice girders shall only be lifted during installation using purpose-built devices to attach the girder to the lifting equipment.

10. In multi-pass excavations, connection details require special attention and protection. Details should be proposed by the Contractor for the Engineer's agreement.

# 307. Forepoling

## 307.1. General

1. Forepoling boards used for tunnelling shall comply with Clause 209.5. They shall be driven from the supporting frame in a slightly upwardly inclined direction as specified by the Designer and should penetrate at least half a set beyond the next excavation cycle.

2. On completion of the excavation cycle the next supporting frame is to be installed. During installation of the frame, exposed ground shall be temporarily supported with sprayed concrete where necessary. The boards installed in the previous excavation cycle are then to be wedged tight against the newly installed supporting frame.

3. The next set of poling boards are then installed immediately below the previous set and immediately above the supporting frame by alternately removing the wedges for one board and then driving in the boards in the position that is now exposed. In open configuration, removal of the wedges might not be required. In the event that full penetration is not achieved, the boards can be fully driven to their required location during the excavation cycle.

4. Upon completion of each advance the specified sprayed concrete lining and reinforcement shall be installed.

5. Great care shall be taken to prevent the disturbance of face boards and supports in general during the forepoling cycle.

6. Forepoling boards shall be used either in open (staggered) or closed configuration in combination with sprayed concrete.

7. Ribs and arches used for forepoling shall comply with Clause 208.2.

## 308. Explosives

### 308.1. General

1. The Contractor shall use explosives only in circumstances where it is safe to do so having due regard to the safety of persons, third-party property and the safety of the Works. Explosives shall not be used without the agreement of the Engineer. The Contractor may appoint an explosives subcontractor to undertake the work with explosives, with the agreement of the Engineer.

2. All work with explosives shall be undertaken in accordance with current legislation and the recommendations of BS 5607:2017 Code of practice for the safe use of explosives in the construction industry. Extensive general guidance on the use of explosives is set out in clauses 1–10 of BS 5607:2017 with guidance on the safe use of explosives in tunnels and shafts in clause 11. The Contractor shall obtain all necessary licences and consents and shall provide secure storage facilities for all explosives and equipment in accordance with BS 5607 and the requirements of the local police force and the Engineer.

3. Explosives shall be handled and used only by personnel duly authorised by the Contractor. The names and qualifications of such personnel shall be submitted to the Engineer in writing in advance of any possible use of explosives.

4. Blasting operations shall be carried out only under the direction of an experienced Explosives Engineer. The Contractor shall appoint one competent person to be responsible for the security of explosives.

5. At an early stage, in advance of the proposed use of explosives, the Contractor shall notify the Engineer, third parties, statutory authorities and services that have an interest in or are likely to be affected by blasting operations of the general nature of the operation. The Contractor shall subsequently give a minimum of 14 days' notice to the Engineer and others described above of the proposed use of explosives. With this notification the Contractor shall submit to the Engineer a detailed method statement on all aspects of the proposed use of explosives, including the treatment of misfires.

Drill and blast pattern shall be designed by a competent blasting engineer considering the vibration and air-overpressure limitations for any structure within the project or third-party structures. The pattern shall be designed in a way to minimise impact on surrounding rock and approved by the Engineer. The blasting result along with vibration and air-overpressure monitoring shall be reviewed after each blasting and the drill and blast pattern shall be modified based on the result if required with the agreement of the Engineer.

6. Blasting shall be carried out carefully using controlled blasting techniques (pre-splitting, line drilling, smooth blasting or

similar) to avoid loosening or shattering rock beyond the required line of excavation, and loose or shattered rock (where it does not contribute to the stability of the excavation) shall be removed by scaling down or other means before personnel will be permitted to restart operations after blasting. The Contractor shall submit the following information to the Engineer for approval before each stage of blasting

(a) geometry of blasting place
(b) excavation drawing, number and depth of the blasting holes
(c) weight and type of explosives
(d) method of charging each hole
(e) total amount of explosives
(f) the quantity and number of delay detonators
(g) design of blasting circuit
(h) quantity of explosive in kg for each delay (mic)
(i) quantity of explosives in kg for every cubic meter of excavated rock (powder factor).

The Contractor shall be given blasting permission after acceptance by the Engineer of the above information.

7. Notices of blasting operations shall be posted on site. Before each firing, the Contractor shall give audible warning, clear the area, and shall take positive measures to prevent personnel from entering the danger area.

8. The Contractor shall monitor the results of blasting closely and, where it is proper to do so, shall propose changes to their blasting operation for the agreement of the Engineer.

9. Under no circumstances shall any holes be charged until completion of all drilling operations at the face.

10. After each blasting operation the tunnel drive shall be sufficiently ventilated to remove any hazardous contaminants and the atmospheric conditions shall be comprehensively monitored prior to personnel accessing the excavated face.

11. Shot-firer/competent authorised person shall visit the blasting area after ventilation to ensure there is no misfire or risk of remaining explosive in the face and it is safe to start the next activities. Misfire shall be dealt with in accordance with requirements of BS 5607:2017.

12. No person shall be allowed to approach the face and no face operation shall commence until the Contractor's authorised person in charge of the operation has given permission.

13. As soon as practicable after blasting and without undue delay, the Contractor shall erect such support as may be necessary to safeguard the excavation and personnel.

14. The shot-firer must keep a record of the number of shots fired, their time of firing, type and weights of explosives used and

the type and number of detonators used, together with a record of the post-blast situation for each and every location.

15. A copy of the record shall be available to the Engineer at the end of every shift on which shots are fired.

16. Relevant legislation and guidance is listed in clause 2 of BS 5607:2017.

17. The Contractor shall observe all the safety requirements of storing, handling, transporting and using explosives and shall keep at their site office a copy of the government regulations regarding the transport and use of explosives and give another copy of the same to the Engineer.

18. The Engineer shall select the magazine location in compliance with all the related laws and regulations of the country and submit the magazine design to the related authorities for approval. Such approval is the prerequisite for implementing the magazine/store design.

19. The Contractor shall post at the site warning and other notices including the instructions on the blasting time. Such notices shall be in all the languages spoken by the site staff and workers.

20. Electrical method of blasting shall not be used during thunderstorms.

## 308.2. Blasting vibrations and air-overpressure

1. For structures in the proximity of blasting, the peak particle velocity shall be measured at the locations immediately adjacent to the structure nearest to the face being blasted or any other location where it is necessary to limit vibration.

2. Vibration monitoring proposals shall be submitted to the Engineer for their agreement.

3. The measurement of peak particle velocity shall be obtained from instruments capable of measuring along three orthogonal axes, one of which shall be aligned parallel to the centre line of the excavation and another shall be vertical. The Contractor is to provide supports for the measuring instrument if so required by the manufacturer's instructions.

4. The measurements of the particle velocities shall be the responsibility of the Contractor. Copies of the readings in an agreed form shall be supplied to the Engineer.
Air-overpressure shall be measured especially if the blasting is close to residential areas where it is necessary to limit the air-overpressure.

5. Equipment for measurement of vibration shall be in accordance with BS 7385-2:1993.

6. Prior to the commencement of blasting in any location, the Contractor shall demonstrate by the use of test firings, or by other means, that neither the peak particle velocities given in BS 7385-2 nor those stated in the Particular Specification will be exceeded.

## 309. Groundwater

### 309.1. General

1. Unless specifically authorised or approved by the Engineer, lowering of the natural groundwater table shall not be permitted.

2. Any proposals to lower the water table shall take account of the risks of causing settlement or of mobilising or increasing contamination. Removal of groundwater shall not cause damage to the Works, nor to third-party property and shall not cause nuisance either by the removal of ground or by settlement.

3. Any impact on other groundwater abstractions shall be assessed, mitigated and monitored in line with the agreement with the holder of the abstraction licence and the regulating authority such as the Environment Agency or local relevant agency.

4. The Contractor shall submit details of dewatering or depressurisation systems and impacts including installation, commissioning, operation, monitoring, alterations and decommissioning to the Engineer for acceptance.

5. Unless detailed on the Drawings or elsewhere specified, the Contractor is responsible for determining the target performance criteria for the dewatering or depressurisation system in order to maintain the stability of the Works and limit impact on affected assets to acceptable levels.

6. The Contractor's working methods and systems shall be designed to control groundwater to permit the construction of shafts, tunnels, breakouts and connections. Methods shall be in accordance with best practice as outlined in CIRIA Report C750 *Groundwater control – design and practice.* The Contractor shall submit proposals to the Engineer, for their acceptance, for controlling and monitoring any dewatering system, including monitoring changes in groundwater level/ pressures and settlement monitoring, prior to commencing installation and running of any dewatering system. The dewatering system shall include a system for identifying ingress of soil material during the dewatering operation.

7. In planning temporary pumping systems, the Contractor shall take due consideration of water quality, pressure, quantity and variations in water levels.

8. The Contractor shall submit the necessary information to the relevant agency to obtain the necessary consents from the appropriate authorities to abstract, recharge and dispose of groundwater. See Clause 507.4.

9. The Contractor shall ensure that upon completion of dewatering or depressurisation, all abstraction and monitoring wells are fully grouted in accordance with the requirements of

the relevant authority and as-built records are provided to the relevant authority and the Engineer.

10. In areas of contaminated land, abandoned mine workings and other possible water-filled voids, the Contractor shall take account of the potential hazards of inundation of the Works.

11. Any temporary works for the control of water shall be removed and the ground reinstated when they are no longer required.

12. Groundwater lowering as a method of ground stabilisation is further considered in Clause 404.

## 309.2. Contamination

1. Groundwater contamination can be caused by

    (a) mobilising or disturbing existing contamination
    (b) introduction of materials that come into contact with the groundwater, for example ground treatments, grouts, soil conditioning related to the construction process.

2. Where materials are added to the ground or come into contact with the groundwater, the Contractor shall ensure that they are assessed and that they are acceptable.

3. The Contractor is responsible to ensure that evidence is submitted to the relevant agency for acceptance and to the Client, that all permits are in place and that the requirements of the relevant legislation are met. This includes, but is not necessarily limited to, Environmental Permitting, Water Framework Directives, Groundwater Directives, Groundwater Daughter Directives and Groundwater source protection zones.

4. Where land or water contamination is known or suspected, redundant drainage and services shall be sealed or removed to prevent them forming preferential pathways for pollution.

5. The Contractor shall submit instrumentation and monitoring proposals to monitor groundwater control and, where necessary, groundwater contamination to the Engineer for approval.

6. Any proposals to lower the water table shall take account of the risks of causing settlement or of mobilising or increasing contamination. Removal of groundwater shall not cause damage to the Works, nor to third-party property and shall not cause nuisance either by the removal of ground or by settlement. Provisions for treatment of contaminated groundwater shall be submitted for the approval of the Engineer in accordance with the prevailing legislation prior to discharge.

# 310. Probing ahead

## 310.1. General

1. Where required, the Contractor shall be responsible for probing ahead of the tunnel face in order to prove or investigate the ground.

2. The selection of plant for probing shall be agreed with the Engineer and shall take into account the probable nature of the ground ahead and its water-bearing capacity. Probe drill rigs shall comply with Clause 302.

3. Probing shall be carried out such that at all times the distance of probed ground ahead of the face is sufficient to allow modification of the method of working ahead of any change in ground conditions. The number of probes, their positions in the face and angles with respect to the tunnel drive shall be governed by the actual ground conditions encountered and the machinery in use, subject to minimum requirements specified on the Drawings.

4. The maximum probed distance ahead of the face shall be governed by the ground conditions and the degree of uncertainty with distance, subject to minimum requirements specified on the Drawings.

5. Should the information obtained from probing indicate that forward ground conditions will require a modification of the method of working, the Contractor shall prepare proposals for the Engineer's agreement.

6. The diameter of probe holes shall be not less than 38 mm.

7. The flush used shall be suitable for the type of ground conditions anticipated and the machinery in use.

8. A trial shall be carried out in advance of tunnelling activities in order to identify the optimum drill bit, drill diameter, flush and percussion rate. These parameters shall be kept constant during drilling to allow direct comparison between probe holes.

9. An accurate and systematic record of probe hole positions (positions in the face and angles with respect to tunnel drives), drill penetration rate, drill parameters (percussion, torque, thrust), flush (colour, percentage return), drilling sounds (loud, quiet, intermittent), water strikes and interpretation of the nature of the ground ahead shall be noted at the time the holes are bored and a copy provided to the Engineer.

10. Full facilities shall be provided for the Engineer to inspect probing work in progress.

# 311. Break-outs from shafts and tunnels

## 311.1. General

1. For the junctions of tunnels and shafts the Contractor shall submit, for the Engineer's agreement, temporary works details and method statement describing the details of the work, with drawings and calculations. They shall take account of relevant information provided by the Engineer and Designer.

2. Tunnel and shaft linings shall not be broken out until all necessary temporary works, including ground treatment, where appropriate, have been completed to ensure the structural and geotechnical stability of the Works.

3. In segmentally lined tunnels and shafts, segments shall not be removed or lining broken open until the temporary supports and/or ground treatment processes have been properly installed. Adequate wedges and packings shall be installed to limit deformation of the lining during breaking out.

4. Where dowels or bicones are used between segments to transfer loads around the opening, the details shall be provided to the Engineer for agreement. Dowels and bicones shall comply with Clause 216. The details shall prevent damage to any gaskets or shall include detailed proposals of how the watertightness is to be maintained.

5. Temporary supports shall not be removed until so agreed by the Engineer and the work for which they are required has been completed and the structure is capable of carrying the imposed loads. Once junction work is commenced it shall be completed as expeditiously as practicable.

6. The break-out shall be of sufficient dimensions to accommodate the permanent structure to be built. Adequate working space shall be provided to enable the Works to be undertaken safely.

7. The break-out for a tunnel shall enable the construction of a stable initial length of an adequate tunnel lining, built within the specified construction tolerances.

## 312. Installation of sheet waterproofing membranes

**312.1. Surface preparation**

1. A project-based installation plan shall be prepared by the installation Contractor for the acceptance of the Engineer before the work commences. Follow-on operations such as reinforcement and concrete works must be coordinated with the requirements of the waterproofing system.

2. Prior to application of the geotextile fleece layer the primary lining shall be surveyed to confirm that it does not encroach into the designed extrados of the secondary lining. Any proposals to rectify areas of the primary lining shall be agreed with the Engineer.

3. The surface shall be prepared in accordance with the manufacturer's instructions. Except where indicated on the Drawings, all fixtures shall be removed from the primary lining prior to application of the geotextile fleece layer. All core holes shall be backfilled with mortar to be flush with the surface of the primary lining.

4. For sheet waterproofing membranes in a drained application, the profile of the substrate (tunnel surface) shall not have any irregularities that exceed a ratio of length to depth of 5:1 and its minimum radius shall be 200 mm. For membranes in an undrained application, the ratio of length to depth shall be 10:1. For membranes based on FPO/TPO the ratio of length to depth shall be a minimum of 10:1 in all applications. The minimum distance the length to depth ratios need be assessed for can be considered as a length of 500 mm. The substrate surface shall be free from protrusions or sharp edges that may lead to membrane puncture. Crushed aggregates of a grain size greater than 10 mm shall not be used.

5. Groundwater penetrating through the primary tunnel lining shall be collected and drained by appropriate measures. This drainage shall be maintained throughout the membrane placing process and shall be so arranged that excess water pressure behind the membrane cannot develop.

**312.2. Geotextile fleece layer**

1. A layer of protective geotextile shall be attached to the substrate by suitable fastenings installed directly through the geotextile fleece. When fixing the geotextile fleece overhead, sufficient fixings shall be installed to ensure the fleece is in close contact with the substrate and is self-supporting. The sheets shall overlap by at least 200 mm if loose laid or 100 mm if spot welded.

### 312.3. Waterproofing membrane

1. The material for the waterproofing membrane shall be in accordance with Clause 211.
2. The applicator shall undergo thorough off-site training approved by the manufacturer of the membrane. The installation contractor shall demonstrate a satisfactory track record for carrying out comparable works for the acceptance of the Engineer.
3. When placing the sheet waterproofing membrane, no other Works shall be carried out in the vicinity that may cause personnel or equipment to come into contact with the sheet waterproofing membrane before it has been protected. If it is likely that excessive dust may be generated in the vicinity of the Works, from vehicle movements and so on, then dust suppression measures shall be put in place.
4. Waterproofing membranes shall not be stored in direct sunlight prior to use. Waterproofing membranes shall be protected from damage at all times.
5. The amount of membrane stored in the tunnel shall not exceed one day's production to minimise the fire load stored underground.
6. The sheet waterproofing membrane shall be fixed to the tunnel structure by means of fastening devices that preserve the integrity of the sheet waterproofing membrane. Sufficient fixings shall be installed to ensure that both the fleece and the membrane are in close contact with the substrate and are self-supporting.
7. All sheet waterproofing membrane overlaps shall be welded in accordance with the membrane manufacturer's instructions.
8. Where waterproofing membrane has been installed in the tunnel invert, it shall be protected from any damage as soon as possible after testing.
9. Circumferential and longitudinal water bars in tunnels and shafts shall be welded to the sheet membrane to create isolated compartments not greater than 12 m in length. These bays should allow for future grout injection.
10. Mobile access platforms shall normally be used to allow fixing of sheet membranes at height. They shall conform with the specific requirements of BS 6164:2019 clause 24.15 and with the general recommendations in clause 24 for plant used underground.
11. If an additional protection sheet is installed this shall be secured to the sheet waterproofing membrane such that the two sheets are in close contact.

**312.4. Construction joints: geotextile fleece**

1. Joints shall be in accordance with the manufacturer's instructions.
2. Joints shall have a minimum overlap as specified in Clause 312.2.

**312.5. Construction joints: sheet waterproofing membrane**

1. Waterproofing sheets are heat welded by hot air or with heating elements. Lap joints with a testing channel are heated wedge welded using welding machines. The part weld width must be at least 15 mm and the testing channel width at least 10 mm. The minimum sheet waterproofing membrane overlap is 60 mm.
2. If protrusions through the membrane are required, they shall be fitted with collars to maintain the watertightness of the system.
3. T-joints are located at minimum 200 mm centres. Star or cross joints are not permitted.
4. The length of material roll shall be procured to enable a complete extrados to be installed as a continuous length. Longitudinal joints shall be avoided.

**312.6. Quality Assurance and control – field trials**

1. Field trials shall be made to demonstrate the capability of the equipment, workmanship, materials and application methods under field conditions.
2. The testing programme shall be started sufficiently early prior to installing the membrane to ensure that the required watertightness can be achieved and allow repetition of the trials should the initial results prove unsatisfactory. All trials and acceptance tests shall be completed satisfactorily by the time installation commences.
3. Prior to construction, trials shall be carried out in order to establish the speed and temperature of joint welding required to achieve welds that are acceptable to the Engineer.
4. If the weld cannot be a lap seam with testing channel at some points – for example, transitions, corners, patches and connections – a minimum 30 mm wide, manually hot air welded lap seam, without a testing channel (full seam), is acceptable. If hand-welded joints are proposed at junctions, then this type of weld shall be pre-tested and agreed with the Engineer.
5. The connection to a profiled waterstop (joint waterbar) shall be made by a minimum 30 mm wide lap seam without a testing channel.
6. Butt jointing of thermoplastic profiled tapes shall be done with a heated blade using a guide mechanism. The resultant bead shall be carefully removed after forming the butt weld. The maximum offset of the stop anchor centres is 2 mm.
7. Elastomer profiled waterbars shall be joined by vulcanising. Site elastomer waterstop joints shall be avoided as far as possible.

### 312.7. Quality Assurance and control – construction testing

1. A visual inspection of the sheet waterproofing membrane shall be carried out as specified in Table 11. Areas where the sheet waterproofing membrane is damaged shall be marked up, repairs carried out and tested in accordance with the manufacturer's instructions.
2. All welded joints shall be tested in accordance with Table 11. Any joints that fail the test and require repair shall be marked with a permanent marker at the time of the test.
3. Repairs and hand-welded joints shall be tested by hand-held vacuum chamber in accordance with Table 11. Where this is not possible, joints shall be inspected visually and using a test tool that penetrates at welding imperfections.

Table 11 Construction testing for sheet waterproofing membrane

| Parameter | Test method | Frequency | Pass/fail criteria |
|---|---|---|---|
| Coverage | Visual | A visual inspection to be carried out continuously while the membrane is applied | 100% coverage |
| Double welded seam joints | DIN 16726 | Every joint | Pressure drop not to be greater than 20% when a 2 bar pressure is applied for 10 min |
| Hand welding and repairs | ASTM D5641-16 (2016) | Every hand-weld and repair | The weld is regarded as leak-tight if the partial vacuum (0.3 bar) builds up rapidly and remains constant over the testing time (30 s) and no bubbles form in the weld. Any arising bubbles are an indication of leaks. |

4. A visual inspection of the fleece shall be carried out. Areas in which the substrate is still visible, or where the fleece is damaged, shall be marked up and an additional layer of fleece applied with a minimum lap of 200 mm around the area.

## 312.8. Defective membrane

1. Where tears, rips or defective joints in the geotextile fleece are noted these shall be repaired with a minimum overlap of 200 mm.
2. Where tears, rips or defective joints in the sheet waterproofing membrane are noted, these shall be repaired in accordance with manufacturer's recommendations. These shall be tested as specified in Clause 312.7.3.
3. Any sheet waterproofing membrane not meeting specified requirements shall be removed and replaced, including any associated water management measures or smoothing layer. The cause of the problem shall be rectified before placing any further elements of the waterproofing membrane, including the drainage and protection layer.

## 312.9. Secondary lining concrete works

1. The placing of secondary lining concrete sequence and processes shall be such that they do not displace or damage the geotextile fleece or sheet waterproofing membrane.
2. Invert protective layers shall be installed immediately after the waterproofing membrane. The waterproofing membrane shall not be exposed to pedestrian or vehicle traffic while unprotected. In case of vehicle traffic there is a general need of a reinforced protection screed. As soon as practicably possible after the membrane has been installed it shall be protected by the construction of the secondary lining.
3. To prevent concrete infiltration, laps on membrane invert protection layers and their connections to the waterproofing membrane shall be heat welded. The protective layer shall be taken at least 300 mm above the invert reinforcement starter bars.

## 313. Installation of sprayed-applied waterproofing membrane

**313.1. Substrate preparation**

1. The risk to human health shall be taken into account when selecting spray-applied waterproofing membrane.

   Prior to application of the waterproofing membrane the primary lining shall be surveyed to confirm that it does not encroach into the designed extrados of the secondary lining. Any proposals to rectify areas of the primary lining shall be agreed with the Engineer.

2. The surface shall be prepared such as to give a suitable finish in accordance with the manufacturer's instructions.

3. Field trials shall be used to determine the most effective method of achieving the required finish. They shall also be used to determine levels of atmospheric contamination to be expected during the application process.

4. Before application of the spray-applied waterproofing membrane, the sprayed concrete surface shall be thoroughly cleaned using compressed air and water (without oil contamination). All surface contamination, such as dust, oil, loose particles and so on shall be removed.

5. If a non-resin-based spray-applied waterproofing membrane is used, the surface shall be damp before application of the spray-applied waterproofing membrane as per the manufacturer's instructions.

6. The substrate surface texture shall be adequately smooth to permit an even continuous spray membrane application with a minimum thickness of 3 mm. Where the surface texture roughness is too excessive to permit full coverage, or the consumption of spray-applied membrane is uneconomical, a regulating layer shall be applied to the sprayed concrete surface. The regulating layer shall consist of sprayed concrete 5–10 mm thick and shall use sand (with a grading of 0–4 mm) as the aggregate. The regulating layer shall not encroach into the designed extrados of the secondary lining.

7. Groundwater ingress shall be pre-sealed by resin injection, fast-setting impermeable mortar, or managed by drainage systems so that there is no active ingress through the substrate surface during membrane application. This drainage shall be maintained throughout the membrane installation works and shall be arranged such that excess water pressure cannot develop behind the membrane.

8. For each area of groundwater ingress, countermeasures shall be submitted to the Engineer for approval.

9. Where a drainage system is used the Contractor must demonstrate the sprayed waterproofing system remains effective where it is applied to or overlaps the drainage system.

**313.2. Application**

1. During membrane application, the environmental conditions local to the spraying shall be in accordance with statutory limits and the membrane manufacturer's instructions. Ventilation should be not less than 0.5 m/s to provide optimal application and curing conditions.
2. When installing the spray-applied waterproofing membrane, no other works shall be carried out in the vicinity that may cause personnel to come into contact with the spray-applied waterproofing membrane before it has sufficiently cured.

**313.3. Equipment**

1. The spray-applied membrane may be applied robotically. Robotic application shall be as approved by the membrane manufacturer.
2. Application of the spray-applied waterproofing membrane shall be in accordance with the recommendations of the manufacturer.
3. During application, particulates and toxics in spray shall be monitored.

**313.4. Construction joints**

1. Where the spray-applied waterproofing membrane is sprayed in alternate bays, or there is an interruption in spraying of more than 6 h, there shall be a minimum overlap of 200 mm with the existing spray-applied waterproofing membrane and the surface shall be cleaned prior to application.
2. Cleaning requirements for construction joints shall be confirmed following site trials, validated by bond strength testing.

**313.5. Defective membrane**

1. Areas of the spray-applied waterproofing membrane that lack uniformity, exhibit lamination or cracking, lack adequate bonding, lack watertightness, or fail to meet the specified strength and toughness requirements shall be regarded as defective membrane. Where an area is deemed defective the section shall be removed, cleaned and resprayed with a minimum overlap of 200 mm from the boundaries of the defect.
2. Any spray-applied waterproofing membrane not meeting specified requirements shall be removed and replaced including any associated water ingress control measures or smoothing layer. The cause of the problem shall be rectified before placing any further spray-applied waterproofing membrane.

**313.6. Secondary lining construction**

1. Prior to secondary lining construction, the membrane shall be inspected for defects, pinholes and 100% coverage.
2. Where a spray-applied waterproofing membrane has been installed then the secondary lining concrete shall not be applied until the spray-applied waterproofing membrane has cured sufficiently to achieve a minimum Shore A hardness of 50.

**3.** As soon as practicably possible after the membrane has been installed, it shall be protected by the construction of the secondary lining.

**313.7. Storage**

Only material sufficient for two days' production shall be stored securely underground at a time. All hazard warnings from the manufacturer's data sheet shall be prominently displayed adjacent to the store. Store location and maximum quantities stored contents shall be added to emergency plans.

## 314. Tunnel boring machines (TBMs) and shields

**314.1. General**

*Note: TBMs are bespoke project-specific items of plant. The capability and functionality required of a TBM depends on the expected conditions of use including ground conditions, rock properties and faulting, groundwater regime, ground support and lining regime, tunnel diameter and alignment along with environmental conditions. Required capability and functionality also depend on proposed operational procedures including services to the TBM, power supply, tunnel ventilation system, spoil removal systems and supply logistics as well as emergency procedures. In addition, the intended method and location of assembly and disassembly can also be relevant. The previous experience of the Contractor and/or Client along with their preferences are also relevant.*

*It is not possible to make allowance for these many variables and unknowns in a generic model specification. The clauses that follow are considered the minimum generic requirements for any TBM and it is up to the user of this Specification to add their specific requirements.*

1. Tunnel boring machinery shall conform in all respects with the relevant versions of BS EN 16191 and airlocks with BS EN 12110. Machinery shall be designed to enable tunnel construction to be undertaken in conformity with BS 6164:2019.

   *Note: The term* machinery *is used in the BS ENs in conformity with CEN standards drafting conventions.*

2. The relevant version of BS EN 16191 or BS EN 12110 shall be that which exists as a harmonised standard, seven calendar days before the date of submission of the tender. Should a more recent version of a standard than the harmonised standard have been published as a prEN, then the prEN shall be the relevant version. All clauses of the standards shall be deemed to be clauses of this Specification.

3. Unless otherwise stated in the Contract, the Contractor shall be fully responsible for the selection, design and supply of tunnelling machines, shields and ancillary equipment. It is the Contractor's responsibility to satisfy themselves as to the suitability of the machinery or shields that they will provide.

4. Remanufactured machines or components shall be acceptable, provided the guidance in the current version of ITAtech Report 5 'Guidelines on rebuilds of machinery for mechanised tunnel excavation' has been adhered to.
   Refurbished machines or components may be considered where appropriate.

5. Transport from factory to site and all assembly and disassembly shall be done taking account of the recommendations in the

current version of *ITAtech Report 12 ITAtech guidelines on services of machinery for mechanized tunnel excavation.*

6. The Contractor shall take into consideration all geological and other relevant information made available to them. The Contractor shall be entitled at their expense to undertake further ground investigation should that be considered necessary.

7. The Contractor shall agree the intended use of the machinery with the manufacturer as required by EN 16191, taking into account information on predicted ground, groundwater, tunnel diameter, tunnel alignment and environmental conditions.

8. Operational procedures shall be agreed between the manufacturer and Contractor including services to the TBM, primary lining erection, spoil removal systems, dust capture/suppression and supply logistics as well as emergency procedures.

9. The Contractor shall agree the intended method and location of assembly, launch, retrieval and disassembly with the manufacturer, taking into account predicted site access and conditions.

10. The Contractor shall ensure the tunnelling machinery selected shall provide adequate settlement control when operating as intended by the manufacturer and in the predicted strata agreed with the manufacturer, and such control shall be sufficient to meet settlement limits where these are defined in the Particular Specification.

11. The excavated diameter shall be sufficient to allow the machinery to be steered to the required tolerances and to accommodate the Contractor's primary ground support provisions and still achieve the internal diameter specified in the Particular Specification.

12. Machinery for intended use in soft ground conditions shall incorporate all necessary equipment for lining erection and spoil conditioning.

13. Machinery for intended use in rock shall incorporate all necessary equipment for ground support.

14. Machinery shall incorporate all provisions deemed necessary by the Contractor to allow safe access within the excavation chamber and/or cutterhead for maintenance and inspection purposes. This can include mechanical face support, airlocks and provision for work in compressed air and so on.

## 314.2. Machine characteristics

1. All tunnelling machines shall be designed with adequate safety margins for the anticipated duty and manufactured to comply with all relevant safety standards. The design and layout of the TBM including control and other workstations shall be carefully considered to provide good ergonomics, visibility, safe access and a safe working environment.

2. Provision shall be made for maintenance, including handling heavy components. Components over 25 kg in weight shall be handled mechanically. Individual replacement of major components such as electric motors, hydraulic pumps and motors, propulsion and face rams shall be possible.

3. The design of the machinery shall include provision for the safe replacement of excavation tools during the tunnel drive. The Contractor shall state in the tender the intended means of ground support to meet this requirement.

4. The external diameter of the TBM or shield shall be designed to produce minimum overbreak and the least necessary clearance for the proper construction of the Works. Design shall take into account the horizontal and vertical alignment to be negotiated. Provision shall be made to limit and correct roll of the machine.

5. Where required by the Specification, provision shall be made in the construction of the TBM for probing and ground treatment ahead of the face. The specific requirements in terms of number, location, length and orientation of holes, pressures, materials, prevention of ingress, performance criteria to be achieved and so on shall be set out in the Particular Specification.

6. All machines and shields except those used exclusively for pipe jacking shall be self-propelled. Where propulsion is by means of hydraulic rams thrusting off previously constructed segmental lining, ram shoes and facings shall be designed to distribute the thrust without causing damage to the constructed lining. Ram shoe pads shall be adequately secured. Propulsion rams shall be capable of operating individually or collectively in any combination. They shall permit the insertion of a key closing segment, if used, in the designed location.

7. Where open shields are to be equipped with face rams for the support of the excavated face, they shall be capable of operating individually or collectively in any combination. Face rams shall be designed to accommodate the loads necessary to make the face secure. The operation of face rams shall be interlinked with that of the propulsion rams.

8. Arrangements for extraction, transport and disposal of spoil shall be appropriate for the material and the quantity to be handled.

9. Where backup sledges or trailers towed by the TBM or shield are used, they shall be designed such that they do not damage the permanent or temporary lining.

10. Segment handling systems shall be capable of transferring segments safely from the point of delivery to the TBM, through the segment path and then picking up and placing them safely and accurately all without damage to any part of a segment. Robotic handling, preparation and erection of segments is permissible provided appropriate guarding is

provided to exclude all persons from the area of movement of any robot or segment being moved automatically.

11. The segment path shall be considered a place where persons may be present at any time and all control of movement on and around the segment path shall take account of this.

12. There shall be sufficient instrumentation and monitoring equipment on the machinery that a real-time check for over-excavation compared with previous excavation cycles can be made at any time.

## 314.3. Guidance

1. A guidance system shall be installed on the tunnelling machine. As a minimum, it shall have a display that shows the position and attitude of the machine relative to the design alignment. That information shall be displayed at the main TBM control station at all times. A permanent record shall be made of all information recorded and displayed.

2. Where required by the Specification, the guidance system shall include information on surface and subsurface features.

3. In respect of TBM power supply, the guidance system shall be considered an essential service in terms of clause 4.5.6 of BS EN 16191.

4. The guidance system shall be error tolerant and be capable of generating, recording and displaying error messages with alarms as required by this Specification.

5. The guidance system shall store data securely, including remote backup even when power is lost and be capable of withstanding malicious external attack. It shall be of sufficient capacity to store all project records.

6. The guidance system shall operate reliably in the tunnel environment. The operational limits on dust, humidity, vibration, temperature and so on shall be set out in the specification. Software to interrogate and download the data at any time, in a form compatible with normal business software, shall be provided.

7. A clear space for the guidance system and associated optical survey checks shall be maintained at all times. Any laser beam shall not be positioned along and within a walkway envelope at a height between 1.5 m and 2.2 m above walkway level.

8. An electronic interlock shall halt, with a timer countdown system, forward movement of the TBM when the guidance system becomes faulty or stops working. A password-controlled override shall be provided.

9. An independent check shall be carried out on the location and attitude of the TBM relative to the designed alignment at intervals as agreed with the Engineer.

10. The tolerances on driven alignment along with checks on alignment shall be in accordance to those in section 328.1

Table 14 *Overall tolerances*, unless otherwise detailed within the Specification.

11. Where required by the Particular Specification, the Contractor shall train the Project Manager's staff to interrogate the system or alternatively employ an independent professional specialist to carry out survey checks periodically.

*Note: Clauses setting out additional capability of the system such as forward prediction of TBM alignment or deviation from alignment tolerances, trajectory for returning to the required alignment, frequency of the recording, accuracy and precision of data, locations at which information shall be displayed along with details of means of access to recorded information shall be added by the user of this Specification.*

**314.4. Fire protection**

1. The TBM shall have a fire detection, alarm and protection package covering machinery and electrical installations that meets the requirements of BS EN 16191 and BS 6164. The Contractor and the TBM manufacturer shall jointly agree on any additional fire protection measures required.
2. The extinguishing media shall be suitable for the classes of fire likely to occur.
3. Electrical cabinets and enclosures, including those in the TBM control station, shall be fitted with fire suppression systems.
4. Hand-held extinguishers, where provided, shall have colour-coded covers.
5. Potentially flammable material including conveyor belting, cable insulation and hydraulic hoses shall conform with BS 6164:2019 clause 13.1.11.
6. A water spray barrier shall be provided at the rear of the back up system. Additional barriers may be provided at the Contractor's discretion.
7. Essential services shall be fire-hardened so that in the event of a fire they remain operable for a period of 2 h. The essential services are

   (*a*) emergency power supplies
   (*b*) all fire suppression systems
   (*c*) TBM emergency lighting
   (*d*) environmental monitoring systems
   (*e*) all controls and tunnel communications.

8. The air supply and control systems to the man lock shall remain operable in all emergencies including a fire in accordance with BS EN 16191, BS EN 12110 and the BTS 'Guidance on good practice for work in compressed air'.

**314.5. Contractor's submission**

1. The Contractor shall, as soon as possible after appointment, prepare and agree with the Engineer a programme for the design, manufacture, inspection, testing, transport, erection and commissioning of the tunnelling machinery.

2. The Contractor shall, as soon as possible after appointment, prepare and agree with the Engineer a series of risk assessments and method statements covering the safe undertaking of the tunnelling operations.

   Tunnelling operations shall be deemed to extend to assembly, launch, use and foreseeable misuse, maintenance, retrieval and disassembly of tunnelling machinery.

3. In particular, the risk assessments and method statements shall cover operation and maintenance of the tunnelling machinery, including but not limited to ground excavation and settlement control, alignment control, groundwater control, grouting around the lining, forward probing and ground stabilisation, face entry procedures, handling, installation and utilisation of consumables, provision of a safe and healthy tunnel environment, transport of materials in and out of the tunnel, crew welfare and transport provisions, fire risk mitigation, emergency procedures.

4. The following aspects shall be addressed by separate documents

   (*a*) hyperbaric operations
   (*b*) assembly, disassembly maintenance and servicing of the tunnel machinery
   (*c*) atmospheric monitoring, ventilation and control of atmospheric contamination including gaseous contaminants, dust, particulates and heat
   (*d*) occupational health and welfare.

5. Mitigation of risk of exposure to atmospheric contaminants shall be undertaken in conformity with BS 6164:2019 clauses 15 and 16. A two-stage alarm system shall be adopted.

6. All monitoring data shall be recorded and shall be made available to the Engineer in real time.

7. The Contractor shall provide a basis of design report for the TBM for the information of the Engineer prior to commencing fabrication of the TBM. The report shall include all design assumptions, loading conditions, elements of the TBM and backup to be covered in the calculations, the method(s) of analysis, equipment to be towed behind the backup equipment. The report shall include

   (*a*) a summary of the Contractor's TBM specification

(b) confirmation of the lead time for the supply of a replacement main bearing

(c) details of the main bearing sealing system, if it has previously been used, the tunnel drive lengths and the pressures encountered.

Where required in the Particular Specification, an assessment of the main bearing shall be undertaken in accordance with the guidance in ITAtech Report 1 *ITAtech Guidelines on standard indication of load cases for calculation of rating life (L10) of TBM main bearings.*

**314.6. Quality inspection and testing**

1. The Contractor shall be responsible for the quality of materials used or present within the machinery and must ensure that all materials used or present are adequate for the task they are to perform.

2. Within 12 weeks of Contract award the Contractor shall prepare a detailed design and manufacturing plan and programme for the machinery along with a detailed quality plan covering

   (a) the manufacturer's management procedures for design, materials used, fabrication, commissioning and performance testing

   (b) independent design checks, inspection and similar tests

   (c) specification for the TBM including a list of design codes, standards and so on with which the manufacturer shall comply

   (d) quality assurance and audit practices undertaken by the manufacturer

   (e) details of the records to be kept and procedures for access to them during the life of the project.

3. The Engineer shall be permitted at any stage during manufacture to undertake any inspections, examinations or tests on or off the manufacturer's premises, and at their expense, they consider necessary to ensure the tunnelling machinery meets the requirements of this Specification. Such inspections, examination or testing shall not release the Contractor from any obligation as specified under the Contract.

4. When tests are to be made at the premises of the manufacturer or elsewhere, except where otherwise specified, the Contractor shall provide such assistance as may be necessary to carry out such tests efficiently.

5. New, remanufactured and reconditioned machinery shall be assembled at the manufacturer's works on completion of fabrication or modification and tested to demonstrate that the machinery as a whole and all individual components operate as specified.

6. An audit of the tunnelling machinery machine against the requirements of EN 16191 and EN 12110 shall be undertaken by the Engineer in conjunction with the Contractor prior to the machine leaving the manufacturer's works.

7. Test running will also be required at site following assembly but prior to commencement of tunnel driving. The Contractor shall prepare a schedule of tests to demonstrate the performance of the TBM and on completion shall pass a copy of the test results to the Engineer prior to tunnel driving commencing.

*Note: The user of the specification shall add any specific test or inspection requirements.*

### 314.7. Personnel and training

1. The Contractor shall prepare a management plan setting out the roles, responsibilities and reporting structures for those who are responsible for driving, maintaining and controlling the machine.

2. The Contractor shall ensure that all key personnel who are responsible for driving, maintaining and controlling the machine have the necessary training and experience and are able to demonstrate competence in the duties that they are required to perform, or carry out their duties under close supervision of such key personnel. Preferably, competence shall be demonstrated within a system of national vocational qualifications. Such training shall include emergency procedures.

3. The Contractor shall provide and maintain a complete list of the names of persons, and their duties, responsible for the operation of the machine who have demonstrated competence to a recognised standard.

### 314.8. Design

1. The load-bearing structural elements of the TBM including the shield, bulkhead(s), main beam, cutterhead and its supports, primary conveyor support structure, lining element handling and erection system, thrust/propulsion/gripper system and back up equipment gantries, as well as any system for towing either backup gantries or ancillary equipment in the tunnel shall be designed using finite element methods (FEM) or other structural analysis software and verified by alternative methods.

2. The TBM design shall take account of available working space and loading conditions applicable during TBM assembly, launch, operation, retrieval and dismantling.

3. The design of structural elements detailed in Clause 314.8.1 shall be subject to an independent design check in accordance with BS 6164:2019 clause 6.4. Shields, bulkheads and thrust/propulsion/gripper systems shall always require a category 3 check.

4. The Contractor shall take all necessary steps independently of the TBM manufacturer and their independent checker to be satisfied as to the adequacy of the structural design of the TBM.

5. The Contractor shall submit a report containing design methods, input loadings and worst case scenarios examined along with copies of all design check certificates to the Engineer.

6. The TBM body shall be designed to withstand all foreseeable loads and forces imposed by the ground and groundwater as well as from TBM operations.
Particular consideration shall be given to the loads and forces arising from operations to correct misalignment.

7. The deformation of the TBM body under all loading conditions, including a prolonged stoppage, shall be limited to always permit the lining to be built to the correct geometry.

8. Any temporary supports required to facilitate assembly, launch, retrieval or disassembly and dismantling shall be supplied with the TBM. Their function shall be clearly indicated on them.

9. The cutterhead structure and bearing with its support system shall be rated to absorb the maximum forces envisaged during TBM operation. These forces shall be accounted for in the structural FEM analysis of the shield body structures for circumstances where the full power may be needed, resulting in maximum shove loading and maximum cutterhead torque.

10. The TBM shall be designed to meet the intended method and location of assembly as agreed between the manufacturer and the Contractor taking into account predicted site access and conditions along with any constraints of the launch site as described in the Works Information. The Contractor shall submit a method statement outlining their chosen method for the launch of the TBM.

11. The Contractor shall design the backup equipment to facilitate TBM launch as set out in the Works Information.

12. The TBM shall be designed to meet the intended method and location of disassembly as agreed between the manufacturer and the Contractor taking into account predicted site access and conditions along with any constraints of the disassembly site as described in the Works Information. The Contractor shall submit a method statement outlining their chosen method for the retrieval and disassembly.

13. Transport to site of all TBM components shall be part of the TBM supply.

14. The TBM and the TBM main bearing shall be designed so the main bearing can be removed rearward from the front bulkhead and a new bearing inserted with the minimum disturbance to the TBM configuration.

# 315. Slurry and earth pressure balance machines

### 315.1. General

1. A slurry machine is a tunnel boring machine with a bulkhead located behind the face to form a pressure chamber. Bentonite slurry or other fluid medium is introduced into the chamber under appropriate pressure to equalise ground pressure or groundwater pressure and to be mixed with material excavated by rotary cutter wheel. The resultant slurry is removed by pumping.

2. An earth pressure balance machine is a tunnel boring machine with a pressure bulkhead located behind the face to form a pressure chamber. The excavated material is retained in the pressure chamber under pressure and is extracted by means of a screw mechanism in an operation integrated with excavation. Liquids and additives may be added to the chamber to mix with excavated material ensuring a homogenous and proper application of the pressure to the excavation chamber.

3. Slurry and earth pressure balance machines shall generally comply with Clause 314.

### 315.2. Machine characteristics

1. The machine shall be fitted with a pressure bulkhead capable of withstanding the total pressure envisaged plus an adequate working and safety margin. Where machine size permits, this bulkhead shall include provision for a means of access to the pressure chamber.

2. The pressure control system shall maintain the required pressure on the face at all periods when the machine is advancing and when standing. Control shall be such that the pressure can be adjusted to suit changing face conditions and maintain stability at all times.

3. Pressure sensors capable of working in air, liquid, spoil/liquid and spoil media shall be mounted on the pressure bulkhead or slurry pipework with pressure gauges in the control cabin reading the pressure in the chamber.

4. All tools on the cutterhead shall be of robust and durable construction in order to minimise the need for replacement during the drive. Excavation tools shall be replaceable from the chamber. A system to indicate wear of the tools shall be provided.

5. If the machine requires the application of compressed air to gain access to the pressure chamber, air locks and bulkheads shall comply with the provisions of BS EN 12110. Work in compressed air shall comply with Clause 401.

6. Propulsion rams and shoes shall be designed to take account of the additional forces required to propel the shield forward with the face pressurised. The load from any one ram or combination of rams shall be limited to avoid damage to the lining.

7. The tailskin shall be of a length to ensure that the tail seals are fully engaged on the last ring built after shoving or on a pipe jacking lead pipe.

8. Tail seals shall be designed to withstand the maximum pressure at the tunnel invert plus additional operational pressures from propulsion and grouting, with an adequate safety margin. They shall prevent ingress of water, slurry, grout and other materials into the tunnel. Tail seals are to be replaceable and accessible for maintenance during operations. The type of seal is to be fully compatible with the tunnel lining used.

9. In appropriate situations consideration may be given to providing space on the inbye side of the seals to facilitate the installation of an additional seal in an emergency.

## 315.3. Annular grouting

Grouting shall be by injection through the tailskin unless otherwise agreed with the Engineer. The grouting system shall include measurement of grout injection pressures and volumes.

## 315.4. Spoil removal

1. The excavation and disposal arrangements shall be capable of dealing with the full range of materials expected. Generally, the disposal system shall accommodate material produced by the cutting equipment. A trap may be provided for the retention of pieces of material that would otherwise cause damage to or blockage of the disposal system.

2. Slurry machines shall be provided with means of accurately controlling and adjusting the density and viscosity of the medium supplied to the pressure chamber and introducing additives where required. Pipework, pumps and separation plant shall be designed to accommodate the maximum rate of advance at which the machine will be progressed. The separation plant shall be such that an assessment of the nature and volume of excavated material can be made.

3. Earth pressure balance machines shall be provided with a screw conveyor of sufficient length that the face pressure can be dissipated along its length. Injection points shall be incorporated into the screw conveyor for the introduction of additives.

## 315.5. Instrumentation

1. The monitoring and control instrumentation shall be grouped in suitable panels allowing good visibility and communications within the working area.

2. Instruments and controls shall include but not be limited to the following

   (a) face pressure gauges
   (b) machine position, orientation, inclination and roll
   (c) cutterhead rotation speed, direction, torque and thrust
   (d) cutterhead door aperture status (where fitted)

(*e*) ram thrust pressures, stroke and speed

(*f*) slurry flows, density and pressure (slurry machines)

(*g*) screw rotation speeds and pressure measured along the length at suitable positions and percentage opening of the guillotine at the point where the screw discharges (earth pressure machines)

(*h*) electrical power

(*i*) means of measuring and recording the volume of material excavated per ring of advance

(*j*) TBM advance rates

(*k*) ram extension

(*l*) grout quantity

(*m*) volume of soil conditioning constituent materials including air, water, foaming agents and polymers

(*n*) if required by the Contract, provisions shall be made for data logging of all the above functions.

## 316. Open-faced tunnel boring machines

### 316.1. General

1. An open-faced tunnel boring machine is defined as one in which there is no bulkhead to seal off the face.
2. Open-faced machines shall be constructed to allow compliance with Clauses 301.1 to 301.3 inclusive and Clauses 314.1 to 314.7 inclusive.
3. Where lateral gripper pads are fitted, they shall provide adequate reaction for the forward motion of the machine without imposing excessive loads on tunnel lining, tunnel supports or the surrounding ground.
4. Where required by the ground conditions, the machine shall be equipped with a water spray system, dust shield and dust scrubber system.
5. The cutterhead shall be designed to allow access to the face for the purpose of inspection, testing, sampling and repair. Access dimensions shall be in accordance with the requirements of BS EN 16191.

### 316.2. Unshielded tunnelling machines

1. The machine design shall permit the installation of ground support close to the tunnel face as required by the ground conditions.
2. The design of gripper pads and trailing gear of unshielded machines shall allow installation of full profile steel arch ribs as close as possible of the tunnel face (less than a distance equivalent to the machine's diameter).
3. The machine design shall allow the mounting of rock drills capable of drilling holes as close as possible to the tunnel face at any point on the tunnel circumference, including sidewalls and invert.
4. The design shall also allow for drilling, spiling, probing or ground treatment holes around the top 120° of the tunnel perimeter at angles of not more than 20° to the tunnel axis and extending at least 10 m ahead of the tunnel face.
5. Provision shall be made in the machine design for installing temporary or primary lining, including sprayed concrete and mesh, behind the TBM gripper assembly, except for machines with a small diameter (less than 5 m).

### 316.3. Shielded tunnelling machines

All shielded tunnelling machines shall comply with BS EN 16191.

## 317. Hand shields and mechanised open shields

### 317.1. General

1. Hand shields are tunnel shields with or without working platforms where excavation uses hand tools. Mechanised open shields are similar but with an excavating mechanism mounted within the shield or acting separately within the shield.
2. Use of hand-held excavation tools shall comply with the guidance given in *The Management of Hand–Arm Vibration in Tunnelling. Guide to Good Practice* by the British Tunnelling Society.
3. Hand and mechanised open shields shall comply generally with Clause 314.

### 317.2. Cutter boom machines

1. Cutter boom machines (roadheaders mounted in shields) shall comply in all respects with BS EN 12111 and BS EN 16191.
2. Cutter boom machines shall have adequate total power and cutterhead power and be equipped with appropriate types and numbers of picks, cutters, disks and/or teeth to excavate the ground efficiently. Self-propelled roadheaders shall comply with BS EN 12111:2014. Excavation shall be in accordance with Clause 301.
3. The cutterhead boom shall be capable of operating both transversely and, vertically and in conjunction with the mobility of the machine, shall be able to excavate the face to a neat profile with minimum overbreak to suit the permanent works. Cutters shall be easily replaceable. When mounted in shields, automatic profiling equipment is recommended.
4. The machine shall be provided with dust capture or suppression equipment including water spray arrangements.
5. Where roadheaders are not provided with an integral spoil loading system the Contractor shall provide a mechanised spoil removal system to suit the speed and operation of excavation. Personnel shall not enter the cutting area.

### 317.3. Backhoe machines

1. Shields equipped with backhoe excavating equipment shall comply with appropriate clauses of BS EN 16191.
2. Where backhoes are not provided with an integral, separate, spoil loading system, the method of operation shall be such that personnel access to the cutting and loading area is not required.
3. All excavating devices shall be fitted with fail to safe load holding valves to prevent sudden movement, should any hydraulic hose failure occur.
4. All movements of the excavating device shall require dual button control to prevent any inadvertent movements.

## 318. Tunnelling machines and shield operation

**318.1. General**

1. The Contractor shall take full responsibility for the performance of the tunnelling machines and shields engaged on the Works. They shall employ personnel who are trained in, and have experience of, the type of machine or shield used.

2. The Contractor shall plan their excavation processes, especially at the commencement and completion of drives, in such a way as to prevent ground movement that might adversely affect the permanent works and existing structures. The Contractor shall present a detailed method statement to the Engineer for their agreement. This statement shall include but not be limited to

   (*a*) the proposed method of transportation and erection of tunnelling machines or shields
   (*b*) the proposed method of commencing the tunnel drives until all the ancillary equipment is installed, including any ground treatment or dewatering required; temporary thrust arrangements shall be detailed
   (*c*) the method of determining ground conditions ahead of the face and the assessment of the risk of altered ground conditions
   (*d*) the proposed method for controlling volume loss from tunnelling
   (*e*) the proposed method for reconciling the volume of material excavated with the volumetric rate of advance of the tunnel
   (*f*) the proposed method of junctioning, including dismantling the machines or shields
   (*g*) the method of ensuring that any voids created during the excavation process are adequately grouted as soon as practicable after excavation.

3. Before tunnelling commences, the Contractor shall ensure that all machine systems are operational including all temporary and permanent ground support systems.

**318.2. Excavation**

1. Excavation shall comply with Clause 301.
2. The method used shall ensure correct alignment at all times without imposing excessive loads on the tunnel supports, lining or on the surrounding ground. Careful control of the working face shall be maintained to prevent overbreak and loss of ground.
3. Where linings are constructed behind a shield without a tailskin, excavation shall not commence until a complete ring of lining is available at the erector. Ring erection shall commence immediately after advancement of the shield is complete.
4. When tunnelling under or near existing structures or utility infrastructure, a specific method statement shall be submitted to the Engineer for agreement.

## 319. Pipe jacking

### 319.1. General

1. Pipe jacking is defined as the installation of a tunnel lining by jacking pipes behind a shield, tunnelling machine or auger boring machine. The technique can also be used to install rectangular or other sections.
2. Excavation shall be carried out from within a machine or shield equipped with jacks capable of maintaining and adjusting the alignment and shall comply with Clause 301.
3. Machines and shields shall comply with BS EN 16191, and appropriate parts of Clauses 314 to 317 of this Specification (excepting 314.2.5 and 314.2.10). Operations shall comply with Clause 318 as appropriate.
4. The Contractor shall be responsible for the equipment and systems to provide the forces necessary for the installation of the full pipe string and for the design, provision and introduction of intermediate jacking stations.
5. Appropriate provision shall be made for supporting an exposed excavation face where necessary.
6. The Contractor shall submit method statements for all operations for the agreement of the Engineer before commencement of work. The methods used shall be in accordance with the *Guide to best practice for the installation of pipe jacks and microtunnels* published by the Pipe Jacking Association.
7. Pipe jacking under railway infrastructure shall be done in accordance with the guidance in Network Rail document NR/L2/CIV/044 Issue 4 *Planning, Design and Construction of Undertrack Crossings.*

### 319.2. Thrust and reception pits

1. Thrust and reception pits and shafts shall be designed and constructed to allow the safe operation of plant, equipment and handling of materials and to withstand all loadings imposed by ground pressure, superimposed loads from surface structures and the maximum anticipated thrust forces. Where permanent works accommodate the thrust arrangements, these shall be designed to ensure that the permanent works are not damaged.
2. In all cases the Contractor shall submit their proposals including calculations to the Engineer for their agreement as required.

### 319.3. Operation

1. All key personnel shall be experienced in the pipe jacking process and hold relevant skills accreditation.
2. Before any particular pipe jack length commences, sufficient pipes and, if required, intermediate jacking station assemblies shall be available to ensure continuous operation.
3. The agreement of the Engineer shall be sought for inclusion in the permanent works of repaired pipes.

4. The jacking force applied by the thrust pit jacks or an intermediate jacking station shall not exceed the allowable distributed or deflected design load for any pipe being jacked.

5. Thrust loads shall be transferred to pipes through a thrust ring that shall be sufficiently rigid to ensure even distribution of the load.

6. Changes to line and level shall be gradual. The manufacturer's stated permitted draw or angular deflection on any individual joint shall not be exceeded.

7. Intermediate jacking stations shall be inserted to a predetermined plan. Operation shall commence when loading reaches a predetermined level, which shall be less than the allowable distributed and deflected jacking loads as determined by the manufacturer.

8. To avoid excessive loading, it may be necessary to undertake continuous jacking until completion of the drive. Where this is necessary, the Contractor shall put in place appropriate measures to minimise noise and disturbance.

9. Where necessary, means shall be provided to ensure that the pipeline remains stationary when face balance pressure is maintained and when any jacking rams are retracted.

10. Where required under the Contract or agreed with the Engineer as part of the operational method statement, a lubricating and ground support fluid shall be injected into the annulus between the exterior of the pipe and the ground. This fluid shall be maintained under pressure until completion of the drive. The lubrication injection points shall consist of a minimum of three holes equally spaced around the circumference of the pipe.

11. Where necessary, the lubricant may contain an approved additive to limit water loss.

12. Where the quantities of lubricant injected significantly exceeds the theoretical volumes, this shall be reported to the Engineer.

13. Where required under the Contract, grouting of the pipe annulus shall be carried out on completion.

## 319.4. Packing and sealing

Joint packing material shall be included at each pipe joint and at any jacking station in accordance with the pipe manufacturer's recommendations.

## 319.5. Monitoring and instrumentation

1. The Contractor shall maintain site records of jacking loads, line and level measurements, the distance moved and the relationship between them. In addition, records of the daily reconciliation between spoil excavated and volume of pipe jack built shall be kept. Copies of all records shall be submitted to the Engineer at intervals to be agreed.

2. The jacking force instrumentation shall be calibrated for each drive by the Contractor. The calibration certificate shall be made available to the Engineer.
3. Survey control and guidance shall generally comply with Clause 319.3. Line and level monitoring shall be carried out in conjunction with the pipe deviation angle. Further monitoring shall comply with Clause 329.

## 319.6. Tolerances

1. Pipe jacking shall be carried out in accordance with the alignment tolerances given in Clause 328.
2. Notwithstanding the specified alignment tolerances, the rate of change of direction in any plane, or combination of planes, shall be agreed with the Engineer, taking into account the pipe length, diameter, over-cut, jacking loads and the manufacturer's recommendations.

## 319.7. Microtunnelling

1. The term *microtunnelling* is generally applied to small-diameter tunnels and pipelines installed by pipe jacking methods behind a remotely controlled tunnel boring machine.
2. The microtunnelling machine shall be selected with regard to the ground conditions, length of drive and other relevant factors.
3. Microtunnelling shall comply generally with the provisions of Clause 319.
4. Microtunnelling machines shall comply generally with the provisions of Clauses 314, 315 and 316 as appropriate.

## 320. Jacked box tunnelling

### 320.1. General

1. Jacked box tunnelling is a technique for installing rectangular reinforced concrete box tunnels at shallow depth beneath existing traffic arteries such as railways and highways. The technique avoids the cost and inconvenience of traffic disruption associated with cut-and-cover tunnelling.

2. The principal components of a jacked box tunnel are the reinforced concrete box that may comprise one or more segments, an open-face cellular tunnelling shield fixed to the front end of the leading segment, hydraulic jacks in jacking stations between segments and at the rear end of the trailing segment, anti-drag systems, lubrication and grouting systems and a combined casting and jacking base constructed within a jacking pit.

3. The tunnelling system is designed to install the box tunnel beneath the traffic artery safely, within alignment tolerances and with ground movements controlled so as to keep movements of the overlying infrastructure within acceptable limits. The tunnel is advanced through the ground in typically 150 mm increments, each alternating with equal increments of face excavation. In formulating the design, account is taken of site-specific factors including ground conditions, topography, working space, access and maintenance of the traffic artery.

4. Guidance can be obtained on the use of an integrated jacked box tunnelling system with anti-drag systems in the paper by Allenby and Ropkins (2007).

### 320.2. Design principles

1. Jacked box tunnelling relies on ground that is capable of standing in an open face, supported if necessary by the internal walls and shelves of a cellular tunnelling shield. Unless totally free-standing, the ground must be capable of being penetrated by the shield's perimeter cutting edges, internal walls and shelves without being destabilised. It may be necessary to modify the ground in advance of tunnelling by dewatering, grouting or freezing.

2. The tunnel shall be long enough to avoid construction activities interfering with the traffic artery. A longer tunnel may be appropriate to avoid the need for expensive jacking pit head-wall works.

3. Intermediate jacking stations shall be used when there is insufficient jacking thrust available at the rear jacking station to advance a monolithic tunnel. Jacking thrust shall be transferred through the jacking base into a stable mass of adjacent ground.

4. Anti-drag systems shall normally be used at both the top and bottom of the tunnel. The former is essential to prevent drag-induced movement of the overlying ground, while the latter provides an economical means of maintaining the box segments

on an accurate vertical alignment when they move off the jacking base. Both anti-drag systems, in combination with appropriate lubrication, reduce drag forces and jacking loads.

5. Control of box horizontal alignment shall be achieved by means of side guides constructed on the jacking base and if necessary by the application of eccentric jacking thrust. It is important to ensure the tunnel face is uniformly excavated so as to minimise eccentric shield embedment loads and the risk of misalignment.

## 320.3. Site investigation

A comprehensive site investigation shall be carried out in the area of the proposed tunnel and its associated temporary works. This will identify the nature of the ground and the presence of groundwater. Both short-term and long-term strength parameters for the ground shall be determined. The ability of the ground to stand in an unsupported face and the effect of groundwater on face stability shall be investigated using open trial pits.

## 320.4. Tunnelling shield

1. The tunnelling shield shall be designed to support the overlying ground and infrastructure and to provide safe working conditions for miners and machine operators.
2. In soft ground the tunnelling shield shall be provided with internal walls and shelves, the front of which penetrate the ground and support the face.
3. In hard free-standing ground the shield perimeter shall be provided with a reinforced cutting edge to assist in accurate trimming and minimise over-break.

## 320.5. Jacking system

1. Sufficient jacking capacity shall be provided to overcome the jacking load with a suitable factor of safety. Additional jacking capacity shall be provided where eccentric thrust is required for steerage purposes.
2. The box segments and all components of the jacking system shall be designed to withstand the maximum anticipated jacking thrust with a suitable factor of safety.

## 320.6. Anti-drag systems

1. Anti-drag systems shall be designed to effectively separate the moving box segments from the adjacent ground by interposing a stationary separation layer.
2. They shall be designed to carry the drag loads imparted to them with a suitable factor of safety. They shall normally be anchored at the jacking pit where the load is taken either into the jacking base or into a stable mass of adjacent ground.
3. The top anti-drag system shall be designed to restrict movement of the overlying ground in the direction of tunnelling so that movements of overlying infrastructure are maintained within acceptable limits.

4. The bottom anti-drag system shall be designed to prevent disturbance of the underlying ground and to maintain ground-bearing pressures within allowable limits.

**320.7. Prediction of ground movements**

1. A prediction of ground movements arising from the tunnelling operation shall be made. An empirical method of prediction based on observations of ground movements on previous similar jacked box tunnel projects is acceptable.
2. Settlement predictions shall take into account the ground conditions, box cross-section dimensions and casting tolerances, depth of cover, shield over-cut and travel distance.
3. Predictions of ground movements in the direction of tunnelling shall take account of the drag forces developed between the top anti-drag system and the box segments, the elastic stiffness of the top anti-drag system and any external restraints to movement of the prism of ground overlying the box.

**320.8. Design**

1. Conceptual design of the tunnelling system shall be carried out by the Designer who shall be competent and highly experienced in jacked box tunnelling.
2. Detailed design of tunnelling system components shall be carried out by competent, experienced engineers.
3. All design shall be independently checked.

**320.9. Construction**

1. The Contractor shall submit their proposals for the construction Works, including Drawings, Specifications, method statements and calculations, to the Engineer prior to commencement of construction. In the case of patented or proprietary systems, the Designer, for reasons of confidentiality, may not wish to submit calculations, in which case they shall satisfy the Engineer as to their capability by reference to experience gained on projects of a similar nature and magnitude.
2. When constructing the casting/jacking base, the top surface and all jacking equipment bearing surfaces shall be formed to tight tolerances. This is necessary to avoid local crushing when the box segments move along the jacking base and to provide an accurate launching surface for the box segments. When constructing the shield, box segments and intermediate jacking stations the outer surfaces and jacking equipment bearing surfaces shall be formed to specified tolerances. This minimises shield over-cut and ground settlements and ensures the safe transfer of jacking thrust. Fabricated steel components will be required in the shield, anti-drag system anchorages and jacking system. In view of the high loads carried by the components and the need for dimensional accuracy they shall be fabricated with care to the required tolerances.

**320.10. Tunnelling operations**

1. The Contractor shall submit detailed method statements for all tunnelling operations for the agreement of the Engineer prior to commencement of tunnelling works.
2. Tunnelling operations shall normally be carried out on a continuous basis in order to maintain a stable face and maximise productivity.
3. All key personnel shall be competent and experienced in jacked box tunnelling. Members of the workforce shall hold a relevant Trade Qualification or a National Vocational Qualification or a Plant Operators Certificate.
4. Adequate on-site training shall be given to key personnel and the workforce.
5. Adequate spares and spare machines shall be held on site. In some instances, limited access into the box may necessitate storing spare excavating equipment inside the box.
6. The jacking force applied at any jacking station shall not exceed the designed allowable force to ensure that no part of the box tunnel, shield, jacking equipment and jacking base are over-stressed. Jacking force instrumentation shall be calibrated by the Contractor and the calibration certificate made available to the Engineer.
7. Box alignment shall be monitored prior to each increment of advance and the correct combination of jacks selected at each jacking station to either maintain or correct the alignment as appropriate. It should be noted that the dimensional magnitude of large jacked box tunnels does not permit a rapid steerage response to misalignment.
8. Real-time monitoring of ground surface and infrastructure movements in the vicinity of the tunnelling works shall be carried out before, during and after tunnelling. The depth of shield embedment and the amount of jacking thrust shall be adjusted in response to observed movements. Where appropriate, the surface infrastructure shall be periodically realigned or resurfaced to maintain it within specified positional tolerances.
9. The Contractor shall maintain records of ground movements, face conditions, jacking loads and box alignment against distance moved. Copies of all records shall be made available to the Engineer at intervals to be agreed.
10. Once tunnelling is complete, the box extrados–ground interface shall be systematically grouted. Residual voids at intermediate jacking stations shall then be infilled with reinforced concrete to achieve a monolithic final structure.

## 321. Construction of segmental tunnel lining

### 321.1. General

1. The type of tunnel lining and/or ground support system to be used shall be as specified on the Drawings and in the Particular Specification.
2. The Contractor shall agree with the Engineer all details for the method of construction of linings and support systems including transport, handling and erection.
3. Before erection of each ring of segmental lining, any loose material or other obstructions shall be removed from the ring building area.
4. All faces of all tunnel lining segments shall be thoroughly cleared of foreign matter and debris prior to placing of the segment.
5. Where the lining is to be grouted, the shape of the ring shall be maintained until the ring is stabilised.
6. Erection of segmental linings shall be by purpose-made mechanical system or be manual, aided by mechanical means, in such a way as not to damage the lining. Manual erection will only be permitted by specific agreement with the Engineer after submission of assessments under the Manual Handling Operations Regulations 1992 (as amended). The erection of linings shall follow the agreed method at all times.
7. The method of segment erection in a lining ring shall be capable of fully closing gasketed joints prior to the linings being loaded by external forces.

### 321.2. Erection of bolted/dowelled lining

1. The erection of each ring will normally commence with the invert segments (except in case of universal ring) and proceed by building subsequent segments on alternative sides where possible up to the key or top segment at the predetermined position. Segment positions shall be maintained during erection and after completion of the ring build.
2. Any radial joint bolts shall be tightened at the time each segment is positioned to maintain joint faces in contact and to maintain compression of the gaskets where used.
3. The first segment placed shall be maintained in its correct position and the circumferential joint bolts of all remaining segments shall be located and loosely secured to allow correct formation of the ring shape.
4. Where dowels are used the segment shall be loaded such that the dowels are fully engaged.
5. After completion of the build and before excavation for the subsequent stroke, all circumferential joint bolts shall be tightened. Further retightening of the bolts shall be performed after completion of the subsequent stroke and prior to erection of the next ring.

6. The roll of the lining shall be maintained in accordance with the limitations of bolt hole clearances to ensure full circumferential bolting of the lining to be achieved or as otherwise specified on the Drawings.
7. Where proprietary forms of segment fixings and fastenings are used, methods of erection shall follow the segment manufacturer's recommendations.

## 321.3. Tapered segmental lining

1. Where tapered rings are used they shall be erected in such orientation as may be necessary to produce the specified alignments and grades of the tunnel.
2. The orientation of the taper shall be decided after each excavation cycle prior to erection of the next ring.

## 321.4. Grouting of bolted/dowelled lining

1. The annulus between the segmental lining and the ground shall be grouted immediately after leaving the shield tailskin or as otherwise agreed with the Engineer. Where grouting through the tailskin is being adopted, this shall be concurrent with the TBM advance.
2. Grouting of segmental linings shall be in accordance with Clause 323.
3. No subsequent advance of the TBM shall be allowed until the annular grouting of the last ring built is completed without the approval of the Designer.

## 321.5. Erection of expanded lining

1. Where unbolted lining is erected using an expanding process, the Contractor shall ensure that the lining extrados is in proper contact with the surrounding ground prior to application of the expanding force and that the excavated profile will permit the expansion process.
2. A range of key segment widths shall be maintained by the Contractor.
3. The excavated profile shall be lubricated to reduce skin friction during the expansion process.
4. A joint lubricating compound shall be applied to all wedge faces of each segment to be expanded.
5. Where the length of the key segment is less than the width of other segments comprising the ring, the pockets formed shall be cleaned of all debris to the excavated profile and filled with the concrete grade specified on the Drawings.

## 321.6. Packing

1. Where specified, packings of form and type set out in Clause 215 shall be inserted between the circumferential faces of segmental lining to assist in distributing machine or shield thrust ram forces.

2. Packing shall be used for the correction of line and level and for segmental plane error corrections and also for ring plane alignment correction. Packing shall not in general exceed 6 mm thickness or half the sealing capacity of the gasket, whichever is the lesser.

3. Packings greater than 6 mm required for designed alignment control shall be subject to the agreement of the Engineer.

4. Packings shall not be used in radial joints, unless shown on the Drawings.

5. Packing at any one point in the circumferential joint shall be feathered out to zero in steps of not more than 3 mm.

**321.7. Defective work**

1. Any segments that are damaged or defective prior to erection shall be indelibly marked and removed from site.

2. Any part of the tunnel lining that does not comply with the required tolerances or quality immediately after erection shall be rectified. Advancement of the face shall be suspended until the Contractor's proposals for rectification have received the Engineer's agreement.

3. The Contractor shall submit proposals for the repair of any rings built into the Works for the agreement of the Engineer.

## 322. Segment gaskets

### 322.1. General

1. Gaskets shall be in accordance with Clause 213.
2. Gaskets shall be fitted into the grooves provided in the segment or within the moulds for cast-in gaskets, in the manner recommended by the gasket manufacturer. The gasket dimensions shall be compatible with the groove profile, subject to the specified tolerances.
3. Gaskets shall be fitted to segments before being taken into the tunnel and shall be protected from damage during transport.
4. Care shall be taken to avoid displacing the gaskets during segment handling. No deleterious material shall be permitted in the groove or on the gasket.
5. Hydrophilic and composite (compression/hydrophilic) gaskets shall be protected from the effects of rain or accidental wetting. Segments with hydrophilic or composite gaskets shall not be erected in standing water.

## 323. Grouting

### 323.1. Cavity grouting of segmental lining

1. The term *cavity grouting* shall mean the grouting required to fill the cavities or voids between the excavated profile and the permanent linings of underground works including that due to ground relaxation and any void between permanent and temporary linings. Grout for cavity grouting shall be as described in Clause 214.

2. Primary grouting is the initial cavity grouting that is applied immediately after a unit of lining has been built.

3. Where primary grouting does not completely fill all cavities, secondary grouting shall be carried out.

4. The Contractor shall provide a grouting method statement for the Engineer's agreement. The proposals shall include details and location of the mixing plant and grout pump, mix design and constituents, pumping rates and pressures, injection points, methods of monitoring, recording and controlling the sequence, preventing grout leakage and reconciling the volume of grout placed with the theoretical volume required.

### 323.2. Primary grouting

1. Primary grouting shall be undertaken at a pressure sufficient to place the grout properly and agreed with the Engineer.

2. Primary grouting shall be timed so as to minimise ground movement.

3. For linings erected behind a closed-face TBM, primary grout shall be injected by way of the shield tailskin unless otherwise agreed with the Engineer. In other cases, primary grout shall be injected through grout holes provided in the linings.

4. In segmental linings grouted through grout holes, primary grouting shall proceed in sequence from invert to soffit in such manner that all air and excess water are expelled from the cavity progressively ahead of grouting. Valves shall be connected into the grout holes in order to allow the grout to set under pressure when the grout hose is disconnected. After the grout has set, permanent plugs shall be installed.

5. Any sealing material or device installed at the leading edge of the ring to prevent grout loss shall be removed upon completion of primary grouting.

6. For segmental linings grouted through the tail shield, the Contractor shall propose a grouting sequence for approval by the Engineer.

7. The Contractor shall ensure that grouting pressures do not result in ground heave or overstress or distortion of lining or distortion or damage to gaskets or damage to other structures.

8. Grouting equipment shall be fitted with a pressure gauge and automatic pressure release valves capable of being preset to a specific pressure. Grout pressure is to be measured at the nozzle with a suitable gauge.

9. Primary grouting to segmentally lined shafts constructed by underpinning shall be carried out after the erection of each ring.

10. Grouting shall be carried out at pressures to completely fill the cavity with grout.

11. Where shafts are constructed by sinking as caissons, grouting shall be undertaken on completion of the primary lining and shall be carried out in such a manner that any lubrication fluid is displaced by grout without distortion of the lining.

12. Where the primary void filling is by pea gravel injection, subsequent grouting shall be carried out in stages to the agreement of the Engineer.

13. Grouting of pipe jack tunnels shall be in accordance with Clause 319.3.13.

### 323.3. Secondary grouting

1. Secondary grouting shall be undertaken in selected rings by means of removing grout plugs from the tunnel lining and drilling a hole to the back of the existing grout.

2. Secondary grouting is the regrouting of lining and shall be completed as soon as practicable but within 14 days of the primary grouting or when the face has advanced 50 m from the location of primary grouting whichever first occurs. Where the TBM machine design does not permit this to be achieved, the Contractor shall submit alternative details to the Engineer for agreement. Secondary grouting shall be at a pressure consistent with filling all voids but shall not exceed the design pressures stated in the Particular Specification.

3. The Contractor shall propose the frequency and positions of the secondary grouting locations that shall include grouting through the crown of the lining.

4. Upon completion of grouting, threaded grout plugs shall be fully tightened into the lining.

### 323.4. Cavity grouting of in situ lining

1. The Contractor shall grout all cavities, voids and spaces remaining unfilled outside the in situ concrete lining. Grouting of a section of lining will not be allowed until that section has achieved its design strength.

2. Procedures for cavity grouting of in situ lining to tunnels and shafts constructed with a waterproofing membrane shall be subject to agreement with the Engineer.

3. Grout for cavity grouting shall be as described in Clause 214, except where otherwise agreed by the Engineer, who may direct that large voids be filled with other materials. The grout consistency shall be sufficiently fluid, but not more so, to ensure that the grout flows freely under low pressure ($<100$ kN/m$^2$) into all parts of the space to be filled by way of grout pipes or grout holes provided for the purpose.

4. The injection points shall be provided and used for cavity grouting at an average of at least one per 2.5 linear m of tunnel and more frequently in any areas of excessive over-break. Vent pipes shall be provided extending to the highest points of cavities. The injection points for cavity grouting in arched roofs shall be located within 500 mm of the crown unless otherwise agreed by the Engineer.

5. The Contractor's proposals for the installation of grout pipes shall be submitted to the Engineer for agreement. Grout pipes and grout holes for cavity grouting shall be at least 40 mm internal diameter.

6. Grouting shall be carried out by equipment similar to that used for segmental tunnel grouting. Grouting pressures shall be such as will not damage the Works or any other property.

7. Grout pipes shall not remain within 25 mm of a finished concrete internal surface, and when no longer required all injection holes in concrete linings shall be filled with dry pack mortar to within 25 mm of the finished concrete surface and finally made good.

## 324. Pointing and caulking

### 324.1. Pointing

Segment joints to be pointed shall be cleaned of all grout, dust and deleterious matter so as to leave the recess to be pointed clean and undamaged. In the case of circumferential joints containing packings, cleaning and pointing shall extend at least to the packing or a minimum of 20 mm. The pointing material shall be pressed into the joints so that they are completely filled and then given a steel trowel or brushed finish, as agreed with the Engineer, flush with the inside periphery of the ring.

### 324.2. Caulking

1. Segment joints to be caulked shall be cleaned of all grout, dust and deleterious matter so as to leave the recess to be caulked clean and undamaged. Caulking tools with widths as close as practicable to the widths of the recesses shall be used. Caulking materials shall be forced into the joints so that the full depth is filled. No visible leaks shall remain on completion.
2. Lead caulking shall be used with spheroid graphite cast iron (SGI) segments.

### 324.3. Lead caulking

1. Lead rod used for caulking shall be as close as practicable in width to the width of the recess to be caulked. The lead shall be driven into the recess to fill it completely, forming a continuous solid mass up to the inner surface of the lining.
2. Where both circumferential and radial flanges are machined, the caulking in the radial joints shall be bonded with that of the circumferential joints.
3. Where circumferential joints are unmachined, packing that has been inserted in joints shall be cut out completely and the joint rendered clean prior to caulking. Caulking shall be built up to a depth of 25 mm in the recesses of the unmachined joints. At corners of segments the caulking in the circumferential joints shall be carefully bonded into the caulking of the radial joints by means of block joints built up in the circumferential caulking recess.
4. Lead wool used for caulking shall be compacted by the caulking tools to form a solid mass for the full depth of the joint.

### 324.4. Cementitious cord caulking

Caulking using cementitious cord shall be used with concrete segments and shall be carefully bonded at joint intersections.

## 325. In situ concrete linings other than sprayed concrete

**325.1. General**

1. All surfaces to be in contact with the in situ concrete lining shall be thoroughly cleaned and scaled of all loose or defective material.

2. The surfaces of waterproofing membranes shall be thoroughly cleaned to remove any loose and foreign materials. They shall be cleaned by washing with a stream of air and water, but care shall be taken not to displace the membrane or its fixing and seals.

3. Concrete shall not be placed in still or running water and shall not be subjected to the action of running water until after the concrete has set. Where water flows from surfaces against which the concrete is placed, it shall be excluded from the space to be filled with concrete.

4. All formwork shall be true to form, securely made and supported, and joints shall be sealed to prevent the loss of cement from the mix. Where required, grout pipes shall be incorporated for pressure relief and subsequent grouting.

5. Concreting shall not commence until the formwork has been inspected and agreed with the Engineer.

6. Concrete shall be placed continuously in each length of formwork.

7. Care shall be taken in the case of exposed concrete faces of the tunnel and shaft linings that no irregularity occurs between successive sections.

8. The build-up of water pressure behind uncured linings shall be prevented.

9. The sequence of work within the tunnels or shafts shall be so arranged that no damage occurs to permanent linings. The proposed sequences and methods of operations shall be agreed with the Engineer.

10. Before any concrete is placed for tunnel linings the Contractor shall demonstrate to the Engineer that their concrete mix, equipment and working methods are capable of producing fully compacted concrete to the required surface finish. If required by the Particular Specification, this shall take the form of a trial length.

**325.2. Temperature monitoring of concrete**

1. The concrete temperature at the time of placing shall not exceed 35°C nor be less than 5°C.

2. Where included in the Contract and where directed by the Engineer, the Contractor shall install an array of thermocouples. The Contractor shall monitor the thermocouples to show that the temperature gradient from the core to the face shall at no time exceed 20°C. Results shall be made available to the Engineer.

### 325.3. Transport of concrete

1. Mixed concrete shall be conveyed to its position in the tunnel by pumping, agitator cars or as otherwise agreed with the Engineer. Alternative methods will be required to prove their success in conveying concrete rapidly, without segregation and the loss of materials.
2. Concrete conveying equipment shall be checked by means of site trials prior to general use for its ability to deliver uniform concrete. Slump tests shall be made on samples of concrete taken from the first and last one-tenth of a batch of mixed concrete. If these slumps differ by more than 25 mm, the equipment shall not be approved for use until the condition causing the inconsistency is corrected. Concrete conveying equipment used shall be examined daily for accumulations of hardened concrete or mortar, or for wear of the blades. Where necessary, the uniformity test may be repeated.

### 325.4. Concrete placing equipment

1. Concrete shall be placed by pumping equipment of suitable types, subject to the agreement of the Engineer.
2. Where pumping equipment is used it shall have adequate placing capacity and be capable of delivering the concrete in a continuous uninterrupted flow. The equipment shall incorporate gauges for measuring the pressure in the delivery line and a pressure regulating system. Pumping equipment, storage hoppers and delivery pipelines shall be lubricated at the start of each concreting operation with a batch of cement–sand mortar and shall be thoroughly cleaned at the end of the operation.
3. Concrete placing using pneumatic equipment shall be subject to the agreement of the Engineer.

### 325.5. Placing concrete

1. Concrete shall be placed while still sufficiently plastic for adequate compaction and shall be carefully worked around all reinforcement and embedded fixtures and corners of the formwork.
2. Concrete shall be placed as close as possible to its final position, in continuous near level layers not exceeding 500 mm. Each layer shall be compacted before succeeding layers are placed.
3. Placing equipment shall be operated by experienced operators only. In general, the concrete placing shall continue uninterrupted until the structure is filled over the entire length of the formwork. In the event of equipment breakdown or if for any other unavoidable reason placing is interrupted, the Contractor shall thoroughly compact the concrete to a reasonable level or flat slope while the concrete is plastic. The concrete at the surface of such cold joints shall be cleaned with a high-pressure air water jet before the concrete achieves a primary set, to provide an irregular clean surface free from laitance. Prior to restarting concreting, the surface shall be wetted. The work shall

be so carried out that a sound dense homogeneous structural element is produced.

4. Concrete shall not be subjected to disturbance between 4 h and 24 h after placing.

5. The Contractor shall keep on the site a complete record of the work showing the time, date and location of concrete placed in each part of the Works. This record shall be available for inspection by the Engineer.

## 325.6. Compaction

1. Concrete shall be compacted to produce a dense uniform mass. Except where otherwise agreed with the Engineer, vibration shall be applied continuously and evenly along the work during the placing of concrete in a manner that does not promote segregation of the components and until the expulsion of air has ceased.

2. Unless otherwise agreed, concrete shall be compacted by high-frequency mechanical vibrators. Immersion-type vibrators or heavy-duty formwork vibrators shall be used.

3. Immersion vibrators shall, wherever practicable, be operated in a near vertical position, and the vibrating head shall penetrate and revibrate concrete in the upper portion of the underlying layer. They shall be withdrawn slowly to avoid the formation of voids and shall be carefully positioned to avoid contact of the vibrating head with the formwork.

4. Vibrators shall not be allowed to contact reinforcement or inserts, nor shall they be used as a means of moving concrete along the formwork. The Contractor shall provide standby vibrators during concreting.

5. Particular care shall be taken with the compaction of concrete surrounding water bars to avoid honeycombing and to prevent the displacement of the water bar. Care shall also be taken to avoid displacement of prefixed pipes, blockouts, thermocouples and the like.

6. Where placing concrete for tunnel linings, formwork vibrators shall be used for compacting concrete in the tunnel arch above the highest openings in the formwork. They shall be operated at intervals of not more than 1.2 m behind the advancing slope of the concrete in the shoulders and crown of the arch. The location and operation of the vibrators shall be carefully coordinated with the withdrawal of the discharge line so as to avoid settlement and flow of the concrete from the filled crown.

## 325.7. Curing and protection

1. Immediately after compaction and thereafter for the curing period the concrete shall be protected against harmful effects of weather, rain, rapid temperature changes, frost and from drying out.

2. All concrete should be allowed to cure by methods that will ensure the production of concrete of the specified quality.

3. Curing materials and methods shall be compatible with any subsequent waterproofing.

4. Periods for curing shall be as recommended in Section 6 of BS EN 13670. The Contractor shall agree their proposals with the Engineer.

5. Concrete shall not normally be placed when the temperature at the location of the Works is below, or likely to fall below, 5°C before the section of work can be completed, except in emergencies.

### 325.8. Construction joints

1. Construction joints shall be positioned only where agreed with the Engineer.

2. Formed construction joints shall be formed using purpose-made stop ends. Expanded metal stop ends shall not be used.

3. Unformed construction joints shall be formed using a grout check or similar so that the exposed edge is a crisp true line.

4. Kickers shall be constructed integrally with the structure below.

5. The joint surface shall be either brushed using water to remove laitance and expose the aggregate without disturbing it, treated with retarder and then water-jetted to remove laitance and expose the aggregate without disturbing it or lightly roughened by light chipping or needle-gunning of set concrete. Hacking of set concrete shall not be permitted.

6. Construction joints shall be clean and damp, with no standing water, immediately before wet concrete is placed against them.

### 325.9. Defective work

1. Concrete that is honeycombed, damaged by faulty curing or fails to attain the specified or characteristic strength and concrete work that in any way fails to comply with the Specification will be considered to be defective.

2. Defective work shall be removed and replaced. The methods used for such removal and subsequent reconstruction shall be agreed with the Engineer.

### 325.10. Formwork

1. Before construction commences, the Contractor shall obtain the Engineer's agreement to the general method and system proposed and shall submit detailed Drawings of the formwork to the Engineer for agreement where required by the Particular Specification.

2. All formwork shall be so dimensioned, constructed and securely braced as to prevent displacement.

3. All joints in the formwork and between the formwork and previous work shall be sufficiently tight to prevent loss of liquid from the concrete.

4. Formers for all chases, grooves, recesses and so on shall be securely fixed as part of the formwork. No part of the concrete shall be cut away for any such item, or for any other reason, without the Engineer's agreement.

5. The face of the formwork shall be clean and applied with non-staining release agent. The agent shall not touch reinforcement or items to be embedded and shall not be allowed to collect in the bottom of the formwork or flow onto previously placed concrete.

6. Before any concrete is placed, the Contractor shall examine and clean out the formwork and ensure that the specified reinforcement cover is attained.

7. Formwork shall be eased, struck or removed in such a manner that the structure is not distorted, damaged or overloaded.

8. Except where otherwise agreed, formwork shall not be eased or struck until

   (a) the concrete has attained sufficient strength to support itself in the position cast without deformation or
   (b) a minimum period in line with Section 6 of BS EN 13670 has passed.

9. Where cyclical casting – for example, in situ concrete tunnel lining – striking times may be agreed with the Engineer following criteria determined from trial lengths.

## 325.11. Concrete finishes

### 325.11.1. Formed surfaces

1. Formed concrete finishes shall be as specified on the Drawings with reference to Table 12 below.

Table 12 Formed concrete finishes

| | |
|---|---|
| F1 | No specific requirement |
| F2 | The irregularities in the finish shall be no greater than those obtained from the use of wrought thickness square-edged boards arranged in a uniform pattern. Fins shall be removed and imperfections shall be made good |
| F3 | The resulting finish shall be smooth and of uniform texture and appearance. The formwork lining shall leave no stain on the concrete and shall be so joined and fixed to its backing that it imparts no blemishes. It shall be of the same type and obtained from only one source throughout any one structure. The Contractor shall make good any imperfections in the finish. Internal ties and embedded metal parts shall not be used |

**2.** All formwork joints for F2 and F3 finish shall form a regular pattern.

**325.11.2. Unformed surfaces 3.** Unformed concrete finishes shall be as specified on the Drawings with reference to Table 13 below.

Table 13 Unformed concrete finishes

| | |
|---|---|
| U1: Screeded finish | The concrete shall be levelled and screeded. No further work shall be applied to the surface unless it is a first stage for a wood float or steel trowel finish |
| U2: Wood float | A pre-screeded finish shall be floated with light finish pressure using a wooden float to eliminate surface irregularities |
| U3: Steel trowel finish | A steel trowel finish shall be first wood-floated and then trowelled under firm pressure with a steel float to produce a dense, smooth, uniform surface. The final surface shall be free from trowel marks |

**4.** When required in the Particular Specification and before commencing concreting, the Contractor shall prepare a trial panel to demonstrate that the required surface finish can be achieved by the equipment and methods proposed. The panel shall be filled with the proposed concrete compacted by the method to be used in the work. When agreed with the Engineer, the trial panel shall be retained and will form the benchmark against which all Works concrete shall be prepared.

**5.** When stated in the Particular Specification, each constituent material shall be obtained from a single consistent source. The aggregates shall be free from any impurities that may cause staining. The mix proportions and the grading shall be maintained constant. The same type of material and release agent shall be used in formwork throughout similar exposed areas.

**6.** Release agents shall be selected to leave no stains on the concrete surface and shall be applied evenly.

**7.** Where the concrete surface is to receive waterproofing it shall be in accordance with the waterproofing system manufacturer's recommendations.

**8.** Permanently exposed concrete surfaces shall be protected from rust marks and stains of all kinds.

**9.** After removal of the formwork, no treatment, other than that approved for curing, shall be applied to the concrete until its surfaces have been inspected by the Engineer.

10. Where any surface fails to comply with the Specification in respect of finish, dimensional tolerance or in any other way, the Contractor shall rectify the work as agreed with the Engineer.

11. The Contractor shall be responsible for preventing any damage to the finished concrete surfaces and shall adopt any necessary protective measures to prevent subsequent staining from any cause.

## 325.12. Fixing bar and mesh reinforcement

1. The Contractor may adjust the position of lap joints to fit in with available stock lengths, or construction joints, subject to the Engineer's agreement to the altered positions. The Contractor shall amend the bending schedules, as necessary, to allow for such alterations.

2. The bending of reinforcement bars shall be in accordance with BS 8666 unless otherwise shown on the Drawings and bending schedules. Bars shall be bent cold.

3. Reinforcement shall be free from all mill scale and thoroughly cleaned to remove all loose rust, oil, grease or other harmful matter immediately prior to being placed in position in the Works and maintained thus until finally encased in concrete.

4. All reinforcement shall be accurately placed, securely fixed and adequately maintained in the positions shown on the Drawings. The reinforcement shall be fixed so that the cover specified on the Drawings or in the Particular Specification is achieved, subject to the tolerances specified therein.

5. Reinforcement shall not be rebent on site unless agreed with the Engineer.

6. Spacers and chairs shall comply with BS 7973-1 and be 'Heavy' category as per Table 1 of that standard. Spacer blocks shall be of comparable strength, durability and appearance to the surrounding concrete and shall be factory, produced. Site-produced concrete or mortar cover blocks shall not be used.

7. Spacers and chairs shall ensure that the reinforcement is correctly positioned, be as small as possible consistent with their purpose, and designed so that they will not overturn or be displaced when the concrete is placed. Wire cast in the block for the purpose of tying it to the reinforcement shall be as specified in Clause 325.12.10.

8. Spacers and chairs shall be fixed in accordance with BS 7973-2.

9. Projecting ends of ties or clips shall not encroach into the concrete cover.

10. Tying wires shall be 1.6 mm diameter soft annealed iron wire unless the Particular Specification or Drawings require the use of stainless steel tying wire. Where stainless steel tying wire is required, it shall be 1.2 mm diameter stainless steel wire throughout the structure.

**11.** Concreting shall not commence until the reinforcement has been inspected in accordance with the Inspection and Test Plan.

**325.13. Inspection of completed structure**

The Contractor shall carry out a cover meter survey over all reinforced concrete surfaces within 24 h of removal of formwork. The cover meter survey shall be undertaken on a 500 mm grid over the whole structure. Access for the Engineer to undertake cover meter surveys shall be provided.

# 326. Shafts

## 326.1. General

1. Excavation shall comply with Clause 301 and the relevant clauses of BS 6164.
2. Prior to excavation, the shaft area shall be thoroughly checked for existing pipes, cables or structures and the appropriate measures taken in agreement with the Engineer.

## 326.2. Safety

1. Shafts shall be provided with guard rails and toe boards or rings at least 1.2 m above the surrounding ground level.
2. At all times shafts shall be provided with safe primary and secondary means of access and egress.
3. Should heavy plant and heavy loads have to be located in close proximity to any shaft, the Contractor shall take into account the effects of these loads as well as any ground loads when designing the shaft.
4. The Contractor shall guard against distortion of shaft lining during construction and the possibility of shaft settlement or uplift at an intermediate stage of construction.

## 326.3. Temporary shafts

1. Full details of temporary shafts required by the Contractor's working methods shall be submitted to the Engineer for their agreement. Such shafts shall be adequately sized for all operations required for the execution of the Works.
2. Backfill for temporary working shafts shall comprise material agreed by the Engineer.
3. Where the Contractor wishes to recover temporary shaft linings the structure shall be removed in safe stages as backfilling proceeds, taking care to maintain the safety and structural integrity of the remaining lining. No part of the temporary works shall be left in the ground within 2 m of the designed final surface level.

## 326.4. Construction

1. Shaft sinking shall be carried out by a method suitable for all the particular circumstances of the site including ground parameters, groundwater, depth and final purpose.
2. Where work is done by underpinning, only that ground that may be safely excavated for the installation of one ring or one unit of support shall be carried out. Until that depth is properly secured by permanent or temporary shaft lining, no further excavation shall take place. In the case of pre-formed rings, securing shall include grouting.
3. Where work is done by a caisson operation, a cutting edge shall be fitted to the leading ring. The Contractor's details for bolting caisson rings shall avoid working at height where practicable. The cutting edge shall be maintained at an even level all round during shaft sinking. Jacking arrangements or

kentledge shall be adequate for the work. The Contractor's arrangements shall ensure the stability of any kentledge.

A lubrication space shall be maintained completely filled with the lubricating material around the full shaft periphery during sinking. On completion of sinking the lubricating material shall be displaced with grout.

4. Concrete walls installed by slurry trench or secant piling techniques shall comply with good practice. The Contractor's proposals shall be submitted to the Engineer for their agreement.

5. Where low-pressure compressed air is used to assist shaft construction, air decks and locks shall be designed by the Contractor to give adequate margins of safety against the air pressures to be used in the shaft. Work shall comply with Clause 401. The Contractor's proposals shall be submitted to the Engineer for their agreement.

6. Excavation in rock shall generally be carried out by methods outlined in Clauses 301 and 308. Where explosives are used, full-width shaft covers and blasting mats shall be installed during blasting.

7. Segmental shafts shall be constructed in accordance with the relevant clauses of Clause 321.

8. Packings in segmental shafts shall be in accordance with Clause 215. Caulking and pointing, where specified, shall be in accordance with Clauses 217 and 324 of the Specification. Where segment gaskets are required, they shall be in accordance with Clauses 213 and 322 of the Specification.

9. Full details of temporary works required by the Contractor's method of working for the construction of tunnel or pipe jack eyes in the shaft lining shall be submitted to the Engineer for their agreement.

10. Shaft bases shall be concreted as shown on the Drawings. In the case of temporary works shafts the Contractor shall submit their proposals for the shaft base structure taking account of ground and groundwater forces and sealing the shaft against water entry. Water pressure shall not be allowed to build up under shaft bases until the shaft has sufficient resistance to prevent flotation. The construction sequence shown on the Drawings shall be followed unless agreed with the Engineer.

11. In the event that dewatering, grouting for ground treatment or freezing is employed, the work shall comply with the corresponding Clauses 402, 403 and 404 of this Specification.

## 327. Timber headings

**327.1. General**

1. Timber headings are defined as small tunnels and excavation shall comply with Clause 301.
2. Timber headings are classified as temporary works and shall comply with Clause 303.
3. Timber headings shall be designed and erected in accordance with the requirements of BS 5975 and the relevant British Standards or Eurocodes for structural design.
4. Calculations and Drawings shall be submitted to the Engineer for their agreement, as part of a fully detailed method statement, describing and illustrating the alignment, depth, dimensions of the timber components, stages of the excavation and erection of the timber heading.
5. While headings that are to contain pipes shall be of the minimum size reasonable to allow the pipes to be properly laid, surrounded with concrete or other specified material and then packed, the minimum heading size shall not be less than the minimum size required to undertake the work safely.
6. Where so required, support shall be withdrawn with the Engineer's agreement as the work of packing or backfilling the heading proceeds. Backfill shall be concrete.
7. Where grouting is required, grout pipes shall be left in the top of the heading projecting behind each head tree and the whole grouted solid. Grouting shall be carried out at the end of each shift or after three settings have been packed, whichever is the shorter interval.

## 328. Tolerances for tunnels, shafts and underground works

### 328.1. All tunnels and shafts

1. Tunnels and shafts shall be constructed to the centre lines required by the Contract or subsequently agreed on site with the Engineer.
2. Unless otherwise stated on the Drawings, overall tolerances allowed in respect of the departure of any point on the internal profile of the structure from its established centre line shall be as given in Table 14, measured on completion of the lining construction and grouting. The tolerance includes all building errors.

Table 14 Overall tolerances

| | | |
|---|---|---|
| 1. Segmental lining | Line and level | ±50 mm |
| 2. Sprayed concrete lining | Line and level | ±50 mm |
| 3. Cast-in-situ concrete lining | Line and level | ±20 mm |
| 4. Cast-in-situ hydraulic invert | Line | ±20 mm |
| | Level | ±10mm |
| 5. Pipe jacking | Line | ±50 mm |
| | Level | ±35 mm |
| 6. Shaft | Verticality | 1:300 |
| | Position in plan | ±75 mm |

Where there are specific hydraulic requirements, no backfalls forming localised ponding shall be allowed.

3. Tolerances for survey deviation are not included in the values of Table 14 and shall be as detailed on the Drawings or within the Particular Specification.

### 328.2. Segmental lined tunnels and shafts

1. The maximum lipping between the edges of adjacent segments constituting one ring shall be 5 mm.
2. The maximum stepping between adjacent segments of individual rings shall be 10 mm.
3. The maximum and minimum measured diameters in any one ring shall be within 1% of the theoretical design diameter of the ring measured on completion of ring build and grouting, or such other tolerance stated in the Particular Specification. This tolerance includes all building errors but excludes deformations after ring build and grouting, such as deformations caused by external loads.
4. Ring roll shall be limited to ±150 mm or as detailed on the Drawings.

**328.3. Sprayed concrete lined tunnels and shafts**

1. The maximum lipping between the edges of adjacent sprayed panels shall be 10 mm.
2. The plane of each sprayed ring shall not depart at any point from the plane surface normal to the longitudinal axis by more than 20 mm.
3. The maximum and minimum measured diameters in any one sprayed ring shall be within 1% of the theoretical design diameter of the ring measured on completion of the ring, or such other tolerance stated in the Particular Specification. This tolerance includes all building errors but excludes deformations occurring after completion of the ring.
4. Measurements of the built tunnel geometry shall be made as soon as it is safe to do so. Criteria for safe entry for the purpose of measurements shall be agreed with the Designer and provided to the Engineer for acceptance.

## 329. Control process of underground works

### 329.1. General

1. Underground works in this regard are defined as temporary or permanent tunnels or shafts, either newly built or existing structures, to undergo significant repair/renovation.
2. The Engineer shall produce a Tunnel Management Plan in accordance with the *Joint Code of Practice for Risk Management of Tunnel Works in the UK* in advance of the commencement of underground works.
3. The Contractor shall appoint within their site team an experienced Monitoring Engineer who shall lead the Contractor's monitoring team.
4. The Monitoring Engineer shall submit to the Engineer for agreement a Monitoring Management Plan in advance of the commencement of underground works.
5. A Daily Review Meeting shall be established prior to the commencement of underground works and will continue until the completion of underground works. The frequency of such review meeting may be increased if requested by the Engineer.
6. Assessments to determine potential damage that will occur to existing above ground and subsurface infrastructure shall be carried out through a staged process as follows.

Stage 1. Determine the zone of influence of the planned Works empirically through settlement contours, taking into account the construction methodology proposed and utilising Risk Categories (CIRIA PRJ PR 30) to determine those assets that require further assessment.

Stage 2. Utilising empirical means and conservative assumptions, determine specific areas of concern for a particular asset through assessment of the predicted ground movement and hence potential damage category (Burland *et al.*, 1977) for said area and the mitigation measures necessary (both at source and within the asset) to manage residual risks.

Stage 3. Where Stage 2 assessments conclude an unacceptable level of potential damage or the structure is identified as having a specific structural concern or historical significance, a more refined assessment utilising soil structure interaction and inherent structural stiffness shall be conducted. Predicted movements from this stage shall be adopted for the assessment, and mitigation measures reducing potential damage to acceptable levels shall be defined for the asset prior to tunnelling where required.

The staged process is presented in general terms in CIRIA C796 (2021) Table 4.1.

7. The outcome of the assessments shall determine the type and amount of monitoring that will be required.

8. Early preconstruction instrumentation requirements shall be determined so that baseline measurements can be taken, for an appropriate period, to establish the stability of the monitoring system and any possible effects of any underlying environmental trends that could be attributed to the Works.

## 329.2. Tunnel Management Plan

1. The Tunnel Management Plan shall

    (a) define the roles and responsibilities of the project team
    (b) identify preconstruction risks and mitigation measures
    (c) outline the construction stage assurance procedures.

2. The Tunnel Management Plan shall consider the independence of the engineering and production functions of the project team.

3. The Tunnel Management Plan shall define the permit system enabling the works – that is, the Required Excavation and Support Sequence or Permit to Tunnel. The permit system shall integrate ongoing feedback of engineering data such as excavation parameters and monitoring results and, if necessary, instigate a design change process.

4. A process of face logging or spoil assessment of excavated material shall be defined within the Tunnel Management Plan for every advance.

5. The probing as required by the design shall be a defined process within the Tunnel Management Plan.

6. A Readiness Review process detailing the checks on documentation and plant/equipment/material available to the construction team shall be included within the Tunnel Management Plan.

7. A Lessons Learned process detailing the consideration made of previous health, safety, quality and environmental impacts on projects of a similar nature across the industry, as well as incidents that may arise throughout the duration of the Works, shall be included within the Tunnel Management Plan.

8. A Competency Management process detailing how competency of the key personnel delivering the Works are recorded, managed and maintained shall be included within the Tunnel Management Plan (see Clause 109).

9. The Tunnel Management Plan shall outline the regular processes, procedures and evidence to be produced for ongoing assurance of the underground works.

**329.3. Monitoring of tunnel excavation**

1. The Contractor shall survey, monitor and record tunnel and shaft construction as it proceeds, to form a record of the work.
2. Monitoring shall generally be per unit of advance and include

    (*a*) line and level
    (*b*) cross-sectional accuracy
    (*c*) shift advance and total advance.

3. Where shields and tunnel boring machines are employed, the Contractor shall monitor the attitude of the shield or machine. The information to be recorded in addition to Clause 329.3.2 shall include

    (*a*) square
    (*b*) plumb
    (*c*) roll.

    Where applicable, the following information shall be recorded

    (*a*) face pressure
    (*b*) slurry density, viscosity, level and flow
    (*c*) cutting wheel speed, rotation direction and penetration
    (*d*) screw conveyor speed
    (*e*) volume of excavated material
    (*f*) type, volume and pressure of primary and secondary grouting.

4. Where grouting is carried out, the type, volume and pressure of grout shall be recorded.
5. All information recorded by the Contractor shall be provided to the Engineer on a daily basis unless another interval has been agreed, including access to raw data extracted from any digital monitoring systems as applicable.
6. Where the Contractor considers that any corrective action they may take will exceed the tolerances in the Contract, they shall so inform the Engineer and obtain their agreement.
7. The nature of the excavated material shall be noted and the strata exposed in the tunnel face shall be mapped and recorded where possible.
8. All groundwater ingress shall be recorded and monitored.
9. All atmospheric testing shall be recorded and monitoring for all gases carried out in accordance with BS 6164.
10. The Contractor shall keep copies of all recent face records at the workface for the information of supervisory personnel.

**329.4. Ground movement monitoring**

1. Unless otherwise provided for in the Contract, the Contractor shall monitor and record the effects of tunnel construction at the surface, including all ground movements and the effects on all structures, including the Works. Where specifically requested, the subsurface effects, including movements of the water table, shall also be monitored.

2. Unless otherwise provided for in the Contract, monitoring equipment and instruments shall be provided by the Contractor to enable the response of structures to be determined. Equipment and instruments shall be installed to the manufacturer's instructions and shall be calibrated and tested as appropriate. Monitoring pins and devices shall be securely fixed in position. Due regard shall be given to the construction of the structure to be monitored and the layout of its primary support.

3. Monitoring shall be referenced to stable survey stations located outside the zone of influence of the Works and not subject to ground movement. Such benchmarks and coordinated stations shall be established and agreed with the Engineer before any ground is excavated and before any ground treatment or dewatering takes place. They shall be checked at intervals during the duration of the Works.

4. The Contractor shall observe, record and analyse the readings to establish trends in movement and reconcile movements measured with those predicted. The Contractor shall provide a copy of all recorded results to the Engineer. They shall make available results to the Engineer in accordance with an agreed programme; however, movement greater than predicted shall be reported to the Engineer immediately.

5. The accuracy and precision of the measurement required shall be selected in recognition of the purpose of the monitoring and the level of risk associated with the structure.

6. A trigger strategy defined by predicted levels of movement and movements that have the potential to inflict damage shall be detailed within the Monitoring Management Plan. Trigger breaches will precipitate a pre-defined mitigation action and will be reported to the Monitoring Engineer and Engineer immediately.

7. Any trigger breach shall result in the immediate notification to the Engineer and the convening of a Trigger Review Meeting between the Contractor, Engineer and Client as well as any relevant third parties – for example, asset owners.

8. The stability of the monitoring system and any possible effects of any underlying environmental trends and seasonal change that could be attributed to the Works shall be established in advance of the Works through an appropriate baseline period. The length of the baseline period prior to construction start in the zone of influence shall be agreed with the Engineer.

9. Prior to construction Works commencing, a defect survey shall be carried out of all structures within the zone of influence and a schedule of defects shall be prepared. This schedule shall be agreed by the Contractor and the owner of the structure, or their representative, prior to the start of construction. Existing pipelines, tunnels and services shall be regarded as structures.

10. During the execution of the Works, defects that have been scheduled shall be inspected and monitored as necessary. Defects that arise during the course of the Works shall be recorded. The Contractor shall keep records of such inspections and a copy shall be available to the Engineer.

11. Monitoring of settlement, scheduled defects and defects arising during the course of the Works shall continue at agreed intervals until movements have returned to those recorded during the baseline period.

## 329.5. Tunnel and shaft linings

When constructing segmental lining, the Contractor shall undertake survey checks pertinent to the accurate erection and position of segments during each ring build. The relative attitudes of the lining and the tunnelling machine/shield shall be recorded.

## 329.6. Daily Review Meeting (DRM)

1. The monitoring instrumentation shall be read on a regular basis – as per Drawings and Monitoring Management Plan – and the results made available for a DRM attended by the senior members of the Contractor's and the Engineer's staff. The DRM will consist of the following agenda items for any underground works

(a) review of minutes from previous meeting
(b) health, safety or environmental issues/incidents in the preceding 24 h
(c) quality issues/incidents in the preceding 24 h, including updates on open quality items
(d) review of previous 24 h Works and progress
(e) review of any observations from inspection of the Works
(f) current ground conditions including face logging/probing records
(g) monitoring results including surface, in-tunnel, existing and new structures as necessary
(h) survey results (including line and level for TBM tunnelling)
(i) as-built profiles (including excavation and spray profiles for SCL works)
(j) production testing results and overall trends in material quantity (including grout testing for TBM works)
(k) definition of the Required Excavation and Support Sheet (RESS) or Permit to Tunnel for the next 24 h including any toolbox measures or contingencies

(*l*) definition of any planned/emergency intervention requirements (for TBM tunnelling)

(*m*) confirmation that adequate plant/materials/resources are available to progress with the Works safely

(*n*) outline of the planned Works for the next 24 h.

2. This DRM shall be held daily during the excavation of the tunnels unless otherwise agreed by the Contractor and the Engineer.

3. At the meeting, the Contractor shall present the current results of monitoring of the tunnels, together with trends in these results and comparison with the deformations predicted by the calculations and the triggers defined by the Monitoring Management Plan.

4. All records from these meetings including face or probe logging, quality testing, survey and monitoring results shall be kept and be available for inspection until the conclusion of the Contract.

5. The outcome of the meeting shall be a set of minutes and either the RESS or Permit to Tunnel Sheet (PTS) agreed by the Contractor, Client Representative and the Engineer, which states that tunnelling may continue as proposed, or gives the requirements for modifications to the tunnelling – for example, shorter advances, smaller headings, higher face pressure and annulus grouting around the TBM.

6. If no agreed report is available by a specified time each day, then the tunnel shall be made safe and tunnelling be stopped.

**329.7. Assurance through monitoring**

1. In line with the Monitoring Management Plan a system shall be developed for monitoring movements so that actions can be taken in a timely manner, thereby ensuring that damage to existing buildings and subsurface infrastructure is within calculated predictions.

2. The system used to guide construction shall relate to specific monitoring activities

(*a*) in-tunnel convergence monitoring (SCL)

(*b*) spoil reconciliation (TBM)

(*c*) ground movement monitoring

(*d*) monitoring of adjacent and overlying structures.

3. The predicted values specified in the design documentation shall be used to indicate whether or not there is cause for concern during tunnel construction. To ensure that the response is appropriate for any specific concern, certain procedures shall

be implemented when a trigger limit is exceeded. These are summarised below.

(a) A full review of the lining performance shall be conducted for the relevant tunnel section and checked against the trigger values. This includes checks on the ground/soil conditions, the quality of construction and the monitoring results provided by the Contractor.

(b) A comprehensive review of the trends for monitoring data specific to the area of concern shall be carried out by the Contractor and the Engineer.

(c) The Contractor shall assess the extent to which the deformations comply with the serviceability and extreme limit conditions.

(d) Together with the Engineer, the Contractor shall decide whether changes in the excavation sequences or TBM face pressure are required. This is an interactive process that will determine whether it is safe to proceed with construction or, if there is reasonable cause for concern, the extent to which it is necessary to implement additional measures or emergency procedures. These measures will be included in a new RESS or Permit to Tunnel.

(e) The Contractor and Engineer shall implement a pre-prepared Action Plan, the emergency response to implement contingency measures within timescales outlined in the same document. If there is reasonable cause for concern, it is emphasised that the response must be rapid.

(f) The performance of the tunnel is kept under continuous review until the monitoring data indicate that trends show a stable condition.

4. At least three trigger values shall be established: a green, amber and red limit. The green limit marks the boundary of normal behaviour. The amber marks the boundary of serviceability while the red trigger should be set below the ultimate capacity of the lining. The Contractor's Action Plan should include pre-planned contingency measures that can be taken if a trigger value is exceeded.

5. If a trigger value is reached, first the site team should check that the reading is correct and consistent with the readings from other instruments. If the trigger has been breached, the Engineer is informed immediately. Contingency measures will be instigated, as directed by the Engineer, in accordance with a predefined Action Plan and as decided in a Trigger Review Meeting. The contingency measures are designed to correct any anomalous behaviour.

6. Any materials, plant or labour required for the implementation of contingency measures shall be readily available on site for timely implementation.

**329.8. RESS/PTS – Required Excavation and Support Sheet or Permit to Tunnel Sheet**

1. Based on the design and the evaluation of the results of monitoring, a RESS/PTS will be issued as the outcome of the Daily Review Meeting (DRM) (see Clause 329.6). In the absence of any approved changes, the RESS/PTS will reflect exactly what is shown on the relevant Drawings.
2. The RESS/PTS shall be prepared and endorsed by the Contractor's Site Manager responsible for the tunnelling works, the Designer and the Engineer on site. Unless all the three signatures are obtained, the proposals indicated on the RESS/PTS shall not be implemented.
3. The RESS/PTS shall address but not necessary be limited to the following matters

   (*a*) the tunnel section (chainages) to which the RESS/PTS is applicable
   (*b*) the support to be installed
   (*c*) the excavation sequence
   (*d*) the method of working related to ground support including staging of application of sprayed-concrete layers and lapping of reinforcement
   (*e*) monitoring to be installed in the tunnel section in question
   (*f*) measures to be taken during stoppage of works
   (*g*) other instructions relevant to the tunnel section in question
   (*h*) reference to relevant Drawings
   (*i*) face pressure
   (*j*) soil conditioning
   (*k*) annulus grouting (around TBM)
   (*l*) planned or emergency interventions.

4. A copy of the RESS/PTS will be given to the foreman in charge of the work in the tunnel and shall be kept at the working face.
5. A RESS/PTS is required for every metre of the length of the tunnels.
6. If for any reason the approved design method of working is changed, then this will be reviewed prior to the DRM and, subject to acceptance by the Engineer, a new RESS/PTS will be issued.

**329.9. Contingency measures and emergency procedures**

1. The Contractor shall determine contingency measures to deal with potential hazards that may affect the Works. The Contractor shall submit for approval to the Engineer an Action Plan that shall detail the actions, procedures and contingency measures to be followed in the event that the monitoring system

shows unacceptable levels of deformation/movement if potential hazards occur.

2. Hazards to be addressed include

(*a*) changing ground conditions
(*b*) excessive movement of the linings
(*c*) excessive ground movement
(*d*) excessive settlement of the existing structures
(*e*) unplanned stoppages
(*f*) mechanical excavation plant failure
(*g*) insufficient labour resources
(*h*) failure of services to underground works (air, light, power, etc.)
(*i*) incidents within underground works
(*j*) delay in supply of sprayed concrete (SCL)
(*k*) delay in supply of segments (TBM).

3. In underground construction works, changes tend to be generally progressive with indications of structure or ground behaviour before failure occurs. For this situation a system of hierarchical trigger levels will be appropriate. This allows proportionate response to adverse indications from monitoring.

4. Trigger levels will be based on the results of assessments of at-risk infrastructure. If the assessment indicates that the at-risk infrastructure is unlikely to be able to tolerate the change due to the Works, then triggers will be set based on the levels of change that will be tolerable.

5. There may be some situations where change is less progressive and monitoring may simply be required to give a yes/no response. In these cases, reporting is simple and systems of triggers are not appropriate.

## 330.  Survey and setting out

**330.1. Datum for the Works**

1. The level datum for the Works shall normally be the agreed national datum or as stated in the Contract but should always be selected so that all levels for the project are positive values to avoid confusion or mistakes when some levels are positive and some are negative figures.

   *Note: Usually chosen as 100 m or 1000 m below national datum.*

2. The plan survey shall be conducted to a Works grid established over the area of the Works. Unless otherwise given in the Contract, the orientation of the Works grid shall relate to the national survey grid.

**330.2. Survey benchmarks**

1. The Contractor shall install all level and survey stations required. Such stations shall be of robust construction, protected against damage and the influence of any movement that may arise from the execution of the Works.
2. The Contractor shall check the condition of and resurvey the survey stations at intervals during the progress of the Works.
3. The Contractor shall provide the Engineer with the location and description of all survey stations, the results of surveys and all calculations. Where required, they shall give adequate opportunity for the Engineer to check such stations prior to their utilisation.
4. The degree of accuracy employed in the survey and setting out shall be such as will allow the alignment, levels and dimensions specified for the Works to be achieved.
5. The Contractor shall ensure that all surveying equipment used for the Works is properly maintained and that the performance of the equipment complies with the manufacturer's specification for accuracy.

**The British Tunnelling Society**
ISBN 978-0-7277-6643-4
https://doi.org/10.1680/st.66434.195
Published with permission by Emerald Publishing Limited under the CC BY-NC-ND 4.0 licence, https://creativecommons.org/licenses/by-nc-nd/4.0/

# 4. Ground stabilisation processes

## 401. Compressed air working

### 401.1. General

1. All work shall be carried out in compliance with the Work in Compressed Air Regulations 1996, SI No. 1656.
2. Airlocks and bulkheads shall conform to BS EN 12110.
3. The recommendations in the BTS CAWG *Guidance on good practice for Work in Compressed Air* shall be followed along with the recommendations in BS 6164:2019 Clause 11.
4. High pressure compressed air work shall be carried out in accordance with ITA/BTS CAWG Report 10 *Guidelines for good working practice in high pressure compressed air.*
5. Work in compressed air shall be considered to be temporary works and shall be managed in accordance with Clause 6.4 of BS 6164:2019 in addition to management requirements in the ITA/BTS CAWG Guide.

### 401.2. Submission of information

1. The Contractor shall submit for the Engineer's agreement a method statement naming the person in charge of the work in compressed air and their deputies, if any, along with the Contract Medical Advisor and a description of the type, capacity and arrangement of plant and medical facilities they propose to install, including low-pressure compressed air plant, standby plant, power sources, air cleaning, air cooling plant, communications systems, bulkhead and locking arrangements.
2. The Contractor shall also submit to the Engineer for their review as required by the Contract, full details of the design of the airlocks and working chamber. Where an airlock is a steel pressure vessel, copies of the most recent hydraulic test of the vessel will suffice.

### 401.3. Initial pressurisation

Once the equipment has been installed, the Contractor shall, in normal circumstances, ensure that an independent check on design and installation to verify safety of the installation is carried out and obtain the agreement of the Engineer prior to pressurising any shaft or tunnel. Procedures for functional pressure testing of the workings shall be agreed with the Engineer. If the Contractor has to apply compressed air for safety reasons in an emergency, then the Engineer shall be informed without delay.

### 401.4. Minimisation of leakage

All necessary precautions are to be taken in order to minimise the escape of air through the ground.

### 401.5. Changes in working conditions

Any change in working pressure or sudden or unexpected change in working conditions shall be immediately reported to the Engineer and shall be logged as an incident and reported accordingly.

**401.6. Submission of daily records**

Records of working pressure, air quality, delivery volume and temperature shall be submitted daily to the Engineer. Systems to monitor these parameters in real time shall relay critical data to the Contractor's Site Manager's offices and be equipped with devices that alarm when preset levels are exceeded.

**401.7. Exposure records**

Exposure records for all personnel working in compressed air shall be available to the Engineer on request. On completion of the compressed air works, a full set of exposure records shall be submitted to the Engineer.

A full set of exposure records shall be offered to the Health and Safety Executive for research purposes.

**401.8. Depressurisation of working chamber**

Where the working chamber comprises a length of tunnel, the depressurisation shall be carried out in stages to minimise ground movements and to allow pore water pressure to re-establish itself in the ground surrounding the tunnel. The air pressure in the tunnel shall be reduced at a rate not greater than 0.2 bar per 8 h. Once the air pressure has reached 0.5 bar the Contractor may depressurise to atmospheric pressure at their discretion.

**401.9. Settlement**

A set of levels shall be agreed with the Engineer on relevant settlement points prior to commencement of decompression. Settlement readings shall be retaken at 24 h intervals to ascertain if movement is occurring due to decompression. Settlement readings shall continue daily for a further period of seven days following full decompression of the workings, during which time the compressed air equipment must be maintained in the event that recompression is required.

**401.10. Emergency procedures**

A notice setting out details of emergency procedures shall be provided at the airlock.

**401.11. Storage of materials**

When bagged cement is used, no more than one day's supply shall be stored in the working chamber. Empty bags shall be removed at the end of each shift. Essential timber for emergency ground support shall be stored in the working chamber but hydraulic oil, spare conveyor belting or other flammable materials shall not be stored in the compressed air tunnel or shaft.

## 401.12. Burning and welding

Hot work shall only be carried out under a Permit to Work procedure. Cylinders of oxygen and fuel gas to be used in burning or welding shall be of the smallest practicable size and only equipment approved by the Engineer shall be allowed underground. All cylinders shall be transported in a robust cage and shall be fitted with non-return valves and flashback arrestors at the cylinder end. The cage shall contain a fire extinguisher and a cylinder valve key. The cylinders shall be removed from the working areas immediately after use. Persons carrying out burning or welding in compressed air shall be provided with fire-resistant 'Nomex' overalls or equivalent. A fire watchman with a hose connected to a suitable water supply shall oversee the operations and remain on duty for 30 min thereafter. Spare cylinders shall not be stored underground.

## 402. Grouting for ground stabilisation and groundwater control

**402.1. General**

1. Grouting for ground stabilisation shall mean injecting grout for the safe progress of the Works, the elimination or mitigation of settlement and the reduction of groundwater inflows into the Works. Cementitious grout with suitable additives will be used, followed, where necessary, with chemical grouts as the nature of the ground to be treated and the purpose of the grouting dictates.

2. Grouting shall be carried out only by contractors employing staff and operatives skilled in the work and notified in advance to the Engineer. They shall produce evidence of satisfactory performance on projects where the purpose of the work and extent was comparable.

3. The Contractor shall carry out such trials, additional tests and ground investigation as they deem necessary to formulate their proposals.

4. The Contractor shall take precautions to minimise hydrofracture stress levels within the ground imposed by grouting that might cause damage to structures and/or heave. Where significant stress changes are likely to be imposed, the Contractor shall employ systems to monitor and protect sensitive structures.

5. The Contractor shall take precautions to avoid injected grout entering sewers, drains, granular drainage blankets or other underground structures.

6. The performance of any grouting system shall be monitored by the Contractor in accordance with the Contract, and interpretation of the results agreed with the Engineer.

**402.2. Contractor's proposals**

1. The Contractor shall agree with the Engineer details of the proposed grouting scheme including

   (a) where grouting design is the responsibility of the Contractor, information and case records to support the grouting proposed in respect of its ability to penetrate the strata and its ground enhancement effect

   (b) where grouting design is the responsibility of the Contractor, specific criteria to measure the adequacy, sufficiency or completeness of the ground treatment

   (c) details of the treatment zone and grout injection patterns with respect to the Works and adjacent structures

   (d) details of plant proposed

   (e) method statement and programme, including arrangements for storage of materials, mixing grout, quality control of grout, recording grouting pressures and grout take and tests to prove the efficacy of the grout in the ground, health aspects associated with the materials and grout proposed at all stages of the process and during excavation of treated ground and means of protecting persons from any adverse effects (*f*) an assessment of the environmental impact of the materials and methods proposed

(*g*) an occupational health risk assessment, including methods of risk reduction on all aspects of the grouting operation.

2. Where the grouting contractor is responsible for the design of the grouting, they shall automatically record grouting pressures and flows and produce ongoing assessments of the grout performance in relation to the objectives of the design.

## 402.3. Drilling

1. Any drilling to be undertaken for the grouting works shall be carried out in such manner as to minimise ground disturbance and soil loss. Where drilling or treatment techniques employing air or foam/air are proposed, the issues of escape of air into the ground and disturbance of previously grouted ground shall be agreed with the Engineer.

2. Drill tubes left in the ground after final use shall be flushed out and filled with an approved cementitious grout. Each tube shall be cut off at least 1 m below ground level and the area restored.

## 402.4. Plant

Plant shall be brought to site and maintained in good working order. Batching and mixing plant shall be provided with gauges and equipment that will control accurately the proportions of materials within the required limits and ensure proper mixing and injection of the grout. Gauges shall be checked at the start of each shift. Spares for plant and spare gauges shall be held at site. Current calibration certificates shall be available on site for all electronic measurement equipment.

## 402.5. Disposal of waste

1. The Contractor shall dispose of leakage and wash-out water from injection points and risers in a safe way and shall not allow them to contaminate the site or watercourses or property elsewhere. The Contractor shall take preventative measures to avoid leakage and shall take measures to stop up leakages should they occur. The Contractor shall submit their proposals to the Engineer for their consent.

2. The Contractor shall adopt proper safety precautions to avoid health hazards to all persons, dependent on the nature of the grouts in use.

## 402.6. Records

The Contractor shall keep full and detailed records as are appropriate to the type of treatment being carried out, including direction and full depth of injection pipe, quantities of materials used, time, location and volume of grout injected, volume of grout to waste by leakage and other reasons and the pressure of injection (measured as close to the injection point as possible), both for initial injections and re-injections. Copies of such records shall be given on a daily basis to the Engineer. Continuous automated monitoring of grout pressures and flows shall be made and presented electronically along with summary records. Records of tests carried out on the treated ground shall also be given to the Engineer.

## 403. Ground freezing

### 403.1. General

Freezing for stabilisation and enhancement of the engineering properties of ground shall be carried out only by specialist contractors or subcontractors employing design staff and operatives skilled in the work. They shall produce evidence of satisfactory performance on projects of comparable type and extent and where the purpose of freezing is also comparable. The requirements of BS 6164 in respect of ground freezing shall be complied with.

### 403.2. Process

1. Ground freezing is a process for temporarily supporting the ground during excavation to facilitate the installation of permanent works. It may be used only to prevent water ingress in the case of competent (self-standing when excavated) rock or for strength enhancement, in addition to the prevention of water ingress, in the case of soils.

2. The process comprises two parts

   (a) design, installation, operation and performance of the ground freezing system – that is, freeze tube layout, drilling of freeze tube holes, alignment surveying of the freeze tube holes, insertion and grouting up of the freeze tubes, coolant mains installation, freeze plant specification and monitoring and control of the process, all of which are specialist freeze contractor items

   (b) prediction of the frozen ground behaviour – that is, evaluation of the rock or soil unfrozen and frozen thermal and geo-technical properties, thermal performance evaluation, heave and thaw settlement estimates and determination of the frozen ground structural behaviour during excavation including strength and deformation calculations, all of which are specialist designer items.

### 403.3. Methods

1. Freezing shall be carried out by the more appropriate of

   (a) a closed recirculating coolant system, typically a compressed primary refrigerant removing heat on evaporation from a secondary non-toxic, non-flammable refrigerant through a heat exchanger

   (b) vaporising a non-toxic, non-flammable cryogenic liquid in an open circuit system and exhausting gas to atmosphere.

2. The freezing process shall solidify groundwater in and around the Works and provide adequate ground conditions, without additional measures, for the safe and proper construction of the permanent works and minimisation of settlement.

**403.4. Precautions**

1. The Contractor shall take all proper precautions for the safety of persons and property on site and elsewhere appropriate to the methods of work and materials used. In particular they shall be aware of the possibility of the release of gases and liquids detrimental to health.
2. Vessels and pipes at extreme low temperatures shall be so protected that there is no possibility of accidental contact.
3. The Contractor shall carry out a risk assessment, including identifying methods of risk reduction, on all aspects of the ground freezing proposals including the risk to persons in the tunnel and the risk to persons off-site from exhaust gas emissions.
4. The Contractor shall take all proper precautions to ensure the safety and security of stored cryogenic liquid.

**403.5. Method statement**

1. Where designed by the Contractor, they shall be responsible for the calculation of the total volume of ground to be frozen, the volume to be frozen at any one time, the intensity of freezing and the numbers and location of freeze and monitoring pipes. Calculations, method statement and programme shall be agreed with the Engineer.
2. The method statement shall include a description of

   (a) plant and materials to be used
   (b) drilling and installation of refrigeration pipes
   (c) installation of surface piping system and refrigeration plant
   (d) installation of instrumentation to monitor the freezing process
   (e) maintenance of the frozen ground during underground construction of the permanent work
   (f) removal of the system and demobilisation.

**403.6. Freeze pipes**

Freeze and monitoring pipe installation shall be carried out by rotary drilling techniques using a low-viscosity, non-toxic water-based drilling mud to suit the strata and conditions present in the volume of ground to be treated. The pipes shall be installed to the accuracy of position and alignment necessary to freeze and monitor properly the desired volume of ground. Freeze and monitoring pipes shall be subjected to a pressure test equal to 1.5 times the expected operating pressure to ensure that refrigerant does not leak into the ground.

**403.7. Plant**

1. Plant shall be of adequate capacity to ensure that the volume of ground to be treated will be frozen and maintained in the required state throughout the construction of the Works. Adequate provision shall be made for regular maintenance and spares and additional equipment shall be held for use in the event of equipment or power failure.

2. Plant and pipework shall be fitted with appropriate valves, controls and instruments to ensure safe and proper operation of the system. Valve operating handles shall be insulated.

## 403.8. Monitoring and records

1. The Contractor shall be responsible for monitoring the process during installation and maintenance of the work. Monitoring shall include

   (*a*) the flow of refrigerant within the system so that losses due to leaks will be detected and remedial measures taken
   (*b*) the pressure of refrigerant in the system
   (*c*) the temperatures of refrigerants and ground
   (*d*) vertical and horizontal movements at the surface of the ground and, where required, at buildings, utilities and other structures.

   The method and frequency of monitoring movement shall be agreed with the Engineer.
2. The Contractor shall monitor continuously flow and pressure of the refrigerant and temperature of refrigerant and ground.
3. The results of all observations shall be made available to the Engineer.
4. Excavation in frozen ground shall not commence until the Contractor is satisfied that the freezing operation is complete and maintenance of the freeze is established and the Contractor has received the Engineer's consent. The Contractor shall ensure that their working practices do not interfere with the integrity of the frozen ground. They shall observe and record the strata as excavation proceeds and, where any ground is shown to be not properly frozen, to enable safe excavation, they shall immediately secure the face, stop work and propose suitable remedial works.
5. Maintenance of freeze shall be terminated only after the permanent structure is complete and with the Engineer's agreement. Upon cessation of freezing the system and ground shall be allowed to warm up naturally. Monitoring of ground temperatures and surface horizontal and vertical movement shall be continued for a period prescribed by the Engineer.
6. When no longer required, freeze and monitoring pipes shall be flushed out and filled with an approved cementitious grout. Each pipe shall be cut off 1 m below ground level and the area restored.

## 404. Dewatering

### 404.1. General

1. Where dewatering operations are used they shall be kept to the minimum necessary for the execution of the Works. All work shall be carried out in accordance with CIRIA Report C515 *Groundwater control – design and practice*.
2. Dewatering will not be permitted unless the Contractor can show by approved calculations and in situ tests that the effect of such dewatering will not adversely affect the Works, will not cause settlements exceeding the limits set down in the Contract and will not cause damage to existing properties and structures.
3. Prior to commencement of dewatering, the Contractor shall notify the Environment Agency (Scottish Environment Protection Agency or other regulatory body) and obtain any necessary consents or permits.
4. Construction work shall not commence until the dewatering operation has been proven to be effective at the agreed monitoring locations.

### 404.2. Contractor's proposals

Details of any proposed ground dewatering system shall be agreed with the Engineer before such systems are installed on site.

Proposed details shall include

(a) a dimensioned plan or plans with appropriate cross-sections showing the size, location, depth of each well, predicted drawdown profile and arrangements for disposal of discharged water
(b) depth of filter zone (and any grout seals) at each well
(c) method of drilling or jetting wells, boring diameter and any drilling muds or additives
(d) type of screen and casing
(e) type, size and capacity of pumps
(f) method statement, programme, risk assessment and safety plan
(g) predicted pumping rates
(h) calculations for the predicted drawdown profile and discharge
(i) assessment of settlement and damage risk including mitigation measures to protect sensitive structures/soils – for example, localised recharge, where applicable
(j) arrangements for the measurement and control of water abstraction and the detection of fines or other material that may be drawn into the system
(k) proposed treatment of wells when they are no longer required
(l) any additional ground investigation necessary to provide data for the above
(m) arrangements for duty and standby power supplies

(*n*) arrangements for monitoring the drawdown of the dewatering, including location of monitoring wells and the frequency of monitoring.

**404.3. Drilling and jetting**

1. Any drilling or jetting to install dewatering or observation wells shall be carried out in such a manner as to minimise ground disturbance and soil loss. Drilling muds such as bentonite shall not be permitted, and only environmentally acceptable biodegradable drilling fluids and additives shall be used where prior approval is obtained from the Engineer.
2. Drilling records of the encountered ground shall be maintained and issued to the Engineer.

**404.4. Plant**

1. Plant shall be delivered to site and maintained in good working order. Plant and pipework shall be fitted with appropriate valves, controls and gauges.
2. Each dewatering well shall be capable of individual adjustment and being shut down and isolated from the rest of the system.
3. Appropriate standby equipment and spares shall be maintained on site at all times.

**404.5. Operations**

The Contractor shall take measures to minimise any planned or unplanned interruptions in pumping. Call-out procedures shall be in place to ensure appropriate personnel are available on a continuous 24 h basis during the period of dewatering.

**404.6. Monitoring and records**

1. If existing structures may be affected by dewatering-induced settlements, a building condition survey shall be carried out prior to commencement of dewatering. The Contractor shall monitor ground levels, property and structures for settlement and damage during the period of dewatering and for a period thereafter as specified in the Contract.
2. The Contractor shall determine the extent of the drawdown profile by means of regular monitoring of observation wells installed at appropriate locations

   (*a*) monitoring ground levels, property and structures
   (*b*) reading from observation wells
   (*c*) pumping rates and discharge from each well.

3. The extent of the zone of monitoring shall be determined based on the predicted drawdown profile.
4. The Contractor shall keep full and detailed records of all monitoring carried out. Copies of such records shall be available to the Engineer.

## 405. Compensation grouting

**405.1. General**

Compensation grouting consists of the introduction into the ground of grout layers to compensate for normal tunnelling settlement and to control ground and existing structure movements.

**405.2. Execution**

Grouting shall in general be in accordance with Clauses 402.1 to 402.6.

**405.3. Monitoring**

Monitoring of compensation grouting shall be in accordance with Clause 329.

**405.4. Assessment**

Continual reappraisal of the effects on the ground and structures of compensation grouting shall be carried out by the Contractor and agreed with the Engineer. Amendments shall be made to the grouting procedures to maintain the designed control.

**The British Tunnelling Society**
ISBN 978-0-7277-6643-4
https://doi.org/10.1680/st.66434.209

# 5. Working environment

## 501. Temporary electrical installations

**501.1. General**

1. The Contractor shall be responsible for obtaining an adequate electrical supply for all their site operations.
2. Installations shall comply with BS EN 60204 Safety of machinery. Electrical equipment of machines and BS 7671 Requirements for electrical installations. IET wiring regulations, supplemented but not superseded by the relevant clauses of BS 6164.
3. If so required by the Engineer, the Contractor shall make available a copy of all certificates prepared upon completion of electrical installations and prepared for all required periodic checks.
4. The Contractor shall appoint a competent person to be solely responsible for ensuring the safety of all temporary electrical equipment on site.
5. The Contractor is to comply at all times with the *Electricity at Work Regulations*.

## 502. Ventilation during construction

### 502.1. General

1. Pits, shaft tunnels and headings shall at all times be kept ventilated to maintain an atmosphere fit for respiration and free from oxygen deficiency, potentially explosive or noxious gases and dust, whether present naturally or otherwise. Ventilation shall also be used to maintain a safe working temperature.
2. The Contractor shall take proper precautions to ensure that the Works are kept in a safe and workable condition throughout. In all tunnelling operations the Contractor shall comply with the relevant recommendations of BS 6164 Health and safety in tunnelling in the construction industry. Code of Practice and HSE publication EH40 *Workplace exposure limits*.
3. The Works shall be undertaken in a way that ensures compliance with The Best Practice Guide for *Occupational Exposure to Nitrogen Monoxide in a Tunnel Environment* by the British Tunnelling Society.
4. Where more than one pollutant is present any adverse interaction between them shall be identified and mitigated.
5. In underground workings and in confined spaces, the air breathed by persons shall contain not less than 19% of oxygen by volume.
6. Smoking is forbidden in tunnels, headings, pits or shafts and all confined spaces.
7. In rock excavation all drill holes shall be wet drilled.

### 502.2. Ventilation systems

1. The Contractor shall agree ventilation proposals with the Engineer. Agreement shall not relieve the Contractor of their obligations under the Contract.
2. Proposals shall include but not be limited to the types of fan employed, siting arrangements where appropriate, the power supply and the fan performance data, together with duct characteristics.
3. In forcing systems, fans shall normally be placed on the surface.
4. If booster fans are to be employed by in-line staging, they shall meet the requirements of BS 6164:2019 15.7.6.
5. The inlet to any surface forcing fan shall have unobstructed access to fresh air. It shall not be in the vicinity of a storage site for oil, chemical or diesel drums. The fan shall also be sited so that it cannot draw in internal combustion engine fumes or gas from charging batteries.
6. Provision shall be made for the fan to be run continuously whether persons are within the underground works or not.
7. Where a fan has been stopped and restarted, the condition of the air shall be tested before personnel enter the tunnel. If only forcing surface-mounted fans are employed, the ventilation system should be restarted and run continuously ensuring that any plugs of oxygen-deficient, flammable or noxious mixtures

of gas are flushed out. Care should be taken that workers do not encounter any plugs of these gases on re-entry to the tunnel. The Contractor should take into account that air residence time in long drivages can be several hours and that layered gases of different densities are difficult to disperse, especially where the gradient of the tunnel changes.

8. The outlet of the duct shall be kept as close to the face as is practicable, designed to avoid turbulence and creation of dust and not more than 10 m away.

9. Where dust is being produced by the tunnelling system, exhaust ventilation shall be used to extract such dust from the working area.

10. Tunnelling shall not continue more than 10 m from the shaft or pit unless positive ventilation has been established.

## 502.3. Monitoring

1. Atmospheric monitoring equipment shall be positioned at each working face, inbye of each airlock, and also within 20 m of the tunnel entrance when the tunnel has advanced 250 m or more. Monitors shall also be provided at suitable intervals not exceeding 500 m along the tunnel. Monitoring equipment shall be capable of continuously monitoring the levels of potentially explosive gases, toxic gases and radioactive gases, as appropriate, and the oxygen content. The equipment shall give both visual and audible warning of the presence of potentially explosive, radioactive or toxic gases and where the oxygen content falls below safe working levels. An immediate and effective means of communicating warnings to the surface shall be installed. The atmospheric monitoring system shall be a fixed system supplemented by portable monitoring equipment as necessary, except in small tunnels where the use of portable equipment only shall be permitted at the discretion of the Engineer.

2. Each working shaft and the full length of all tunnels shall be monitored continuously in accordance with Clause 1 of Clause 502.3 for the presence of explosive or noxious gases or lack of oxygen. Records shall be kept of monitoring results. Should the workings be found to contain explosive or noxious gases above the level set out in BS 6164 or HSE guidance EH40 or oxygen content below the level set out in BS 6164, all work shall stop and the Works shall be evacuated until a safe atmosphere is established.

## 502.4. Start-up ventilation

Following an unintended shutdown of the ventilation system, a start-up procedure shall be invoked that will take account of the data from the continuous monitoring system in the tunnel to ensure the atmosphere is safe.

**502.5. Checking and inspection**

1. During each shift, the following checks shall be made

   (a) the fan or fans shall be checked for heat, unusual noise and vibration. The results shall be reported and remedial action taken if required

   (b) the ventilation ducting shall be visually checked for damage and the joints checked for integrity. The results shall be reported and remedial action taken if required

   (c) the atmospheric monitoring system shall be checked at both local and remote stations and the results recorded.

2. The air flow quantities shall be checked at both the face and 20 m from the shaft bottom on a weekly basis. These figures shall be recorded and compared with the calculated flows. Any shortfall shall be made good.

3. The ventilation records shall be maintained and be made available for inspection by the Engineer.

**502.6. Ventilation failure**

1. In the event of ventilation equipment failure all personnel shall be withdrawn from the underground workings.

2. In the event of ventilation equipment failure, where a tunnel boring machine is in use, it shall automatically be stopped and isolated until the ventilation is restored.

**502.7. Ventilation after breakthrough**

After tunnel breakthrough, ventilation facilities designed to ensure safe atmospheric conditions throughout the tunnel system shall be installed.

# 503. Lighting

## 503.1. General

1. Floodlighting on the site surface shall be adequate for the safe operation of the site. It shall be shrouded where necessary to ensure the light is directed to areas within the site and to avoid nuisance.
2. Lighting in the tunnel shall extend the full length and not be less than that required for safe working and access.
3. An alternative source of power and emergency lighting system shall be provided to allow emergency securing operations and evacuation safely in the event of a primary power failure. An adequate number of hand lamps shall be located at key points underground.

All power and lighting wires shall be installed and maintained in optimal conditions of insulation and safety.

All wires shall be firmly fixed on tunnel walls, by means of insulators of adequate design and capacity and their installation shall be made in such a way that wires do not get damaged during tunnel construction.

## 504. Noise and vibration

### 504.1. General

1. The Contractor shall minimise occupational exposure to noise and vibration, the amount of noise emitted to the environment and the environmental vibration levels generated by their work activity. Reference is to be made to the requirements of the Control of Pollution Act 1974, with particular reference to Part III sections 60 and 61, Part III of the Environmental Protection Act 1990, the Control of Noise at Work Regulations 2005 and the Control of Vibration at Work Regulations 2005 in all respects, notwithstanding any liabilities, obligations or restrictions given elsewhere.

2. The Contractor shall follow the recommendations set out in BS 5228 Parts 1 and 2 on control of noise and vibration arising from the Works. Vibration limits shall conform to BS 7385 Part 2 and BS 6472 Parts 1 and 2.

3. The Contractor shall select and utilise methods of working and items of plant and control in the Works so as to minimise noise and vibration levels, including occupational noise and vibration exposure of the workforce, and not to exceed maximum permitted noise and vibration levels specified in the Contract. In particular, the Contractor shall take into account the legislation referring to the exposure levels from hand-held pneumatic tools and comply with the BTS publication *The Management of Hand–Arm Vibration in Tunnelling. Guide to Good Practice*. Where noise and vibration limits are the subject of notices under section 60 or 61 of the Control of Pollution Act 1974 the Contractor shall comply with the requirements of the Specification in addition to those requirements imposed by the sections 60 and 61 Notices.

4. The adherence to any vibration levels specified in the Contract does not relieve the Contractor of their obligations with respect to structural or other property damage, or their obligations under the Control of Pollution Act 1974.

### 504.2. Temporary fencings and barriers

1. Where required the Contractor shall erect and maintain throughout the construction period temporary fencing of appropriate height taking account of the need for this fencing to act as a noise barrier around all working areas. The fencing shall be dismantled and re-erected as the progress of the Works requires.

2. The line of the fencing shall be uniform and the exterior face of the fencing shall be treated with a durable finish. Where required, in order to prevent reflection of noise, the Contractor shall line the inside of fencing with sound-absorbent material with accepted acoustic absorption properties. The material shall be fire- and water-resistant.

3. Local fencing barriers or shelters shall be erected as necessary to shield particular activities, such as those involving the use of pneumatic or hydraulic techniques and all stationary plant. The guidance and advice detailed in BS 5228-1 shall be applied.

## 504.3. Plant and equipment

1. The Contractor shall select and use plant, equipment and working practices that minimise occupational exposure to noise and vibration and minimise emissions of noise and vibration to the environment.
2. All plant shall be properly maintained and relevant service records completed. All plant shall be provided with effective silencers and vibration-dampening devices and shall be operated according to the manufacturer's recommendations in such a manner as to avoid causing any excessive noise emission or vibration. The noise emitted by an item of plant shall not exceed the relevant values quoted in the most stringent of the relevant EU Directive/UK Statutory Instrument and the relevant values quoted in BS 5228-1. All plant operating on the site in intermittent use shall be shut down in the intervening periods between use.

## 504.4. Transport restrictions

On specific contracts it may be necessary to restrict traffic movements at peak times, and to avoid undue disturbance to residents. Truck movements may also be precluded for a period at sites that are located near to schools.

Any specific requirements are given in Clause 106 General provisions and in the Particular Specification.

## 504.5. Noise and vibration monitoring

1. Where monitoring is required, the Contractor shall provide, calibrate and operate according to the manufacturer's recommendations appropriate equipment for monitoring construction noise and vibration throughout the construction period.
2. Noise analysers shall be capable of measuring unattended the equivalent continuous noise level, $L_{Aeq}$, to the Class 1 standard set out in BS EN 61672-1:2013.
3. Vibration measuring systems shall be in accordance with BS 7385.
4. The Contractor shall arrange for adequate standby equipment.

## 504.6. Noise and vibration levels

1. The Contractor shall measure the noise and vibration levels generated by the construction work during working hours throughout the period of construction.
2. The Contractor shall notify the Engineer immediately whenever the specified noise or vibration limit has been exceeded and agree measures to avoid repetition.

3. Any items of plant causing excessive noise or vibration levels shall be removed from the site and substituted with alternative compliant equipment.

4. The Engineer may instruct the Contractor to devise and use an alternative process if a construction method is causing unnecessary disturbance.

## 505. Access and egress

### 505.1. General

The Contractor shall make all arrangements and assume full responsibility for transportation to the site of all construction plant, materials and supplies needed for the proper execution of the Works.

### 505.2. Designated access routes

Where designated access routes are indicated in the Contract, the Contractor shall use no other without the agreement of the Engineer.

### 505.3. Maintenance of routes

1. All public and private highways and roads that are being used by the Contractor's, Subcontractors' or Suppliers' vehicles for the construction of the Works shall be kept clean and free of dirt and mud arising from the Works. The Contractor, unless otherwise provided for in the Contract, shall provide, maintain and use as necessary suitable equipment, including mechanical road sweepers, throughout the course of the Works where and as agreed with the highway authority.

2. The Contractor shall provide, maintain and use mechanical wheel washers and high-pressure hosing facilities at work sites and at such additional locations as required under the Contract.

3. The Contractor shall be responsible for all maintenance in all respects of all site roads.

4. Any area of public highway that is closed because of the Works shall not be reopened until appropriate safety and traffic management measures have been completed and until the Engineer confirms that it is in a suitable condition for use by the public.

5. The Contractor shall protect the public from the Works by secure fencing and gates and shall control access through the gates as required under the Contract.

### 505.4. Access for others

1. The Contractor shall at all times meet the full requirements for access for Fire, Ambulance and other emergency services and maintain liaison with them in that respect.

2. The Contractor shall at all times maintain access for the authorised representatives of utility providers and allow emergency operations to be carried out on any utility or service facilities within the site.

3. The Contractor shall not use public or private rights of way for depositing or storing plant or materials. The Contractor shall maintain those parts of the public or private rights of way not temporarily occupied by the Works in a clean, passable and safe condition at all times.

4. The Contractor shall execute the Works in such a manner that safe pedestrian access, including disabled person access, to all properties is maintained at all times.

5. Unless otherwise provided in the Contract, methods of construction and programming of the Works shall be such that vehicular access to properties affected by the Works is not restricted.

**505.5. Traffic safety and management**

1. When carrying out work on trafficked highways, the Contractor shall comply with the New Roads and StreetWorks Act 1991 and HSE's guidance booklet *Safe use of vehicles on construction sites: A guide for clients, designers, contractors, managers and workers involved with construction transport* so far as it affects personnel who are required to undertake work on highways.

2. Where work is carried out on or adjacent to a trafficked highway the Contractor shall ensure that personnel shall, at all times, wear high-visibility fluorescent garments in accordance with Chapter 8 of the *Traffic Signs Manual*. Garments should comply with BS EN 471.

3. All proposals, details, execution, maintenance, removal and necessary reinstatement associated with traffic safety and management and temporary decking and other temporary structures on, or subways beneath, the highway shall be subject to the approval of the appropriate authorities. The Contractor shall supply all information required for consultation with the appropriate authorities including the local authority, police and other authorities with jurisdiction or interest.

4. The Contractor shall agree a traffic management plan with the Engineer based on consultation and agreement with highway authorities. This shall show the scheme of traffic safety and management measures including the provision of safety zones and traffic signing. The plan shall include the requirements of emergency services for access into and through the site.

5. Fenced storage areas, gantries, loading bays, skips and other temporary structures on the public highway shall be provided and maintained to the conditions of a licence issued by the local authority.

6. All traffic safety and management measures necessitated by the Works shall be fully operational before the Contractor starts any work that affects the public highway.

7. The Contractor shall devise and put into effect traffic management procedures, including appropriate speed limits, within the site including on haul roads and temporary access roads, which are to an equivalent standard to those for a public highway unless directed otherwise by the Engineer.

**505.6. Signing and signalling**

1. The Contractor shall provide suitable entry and exit signs, at the points of access to and from the site, for vehicles and plant engaged on the Works. As far as possible, vehicles and plant shall enter and exit the site in a forwards direction.

2. Unless otherwise specified, the Contractor shall make all necessary arrangements including notices to relevant authorities for the provision, erection, maintenance, repositioning, covering and uncovering and final removal of all traffic signs as the progress of the Works requires.

3. The Contractor shall devise and put into operation traffic management arrangements to separate pedestrian and vehicular traffic. Pedestrian access shall be clearly signed and provided with barriers of adequate strength.

4. The Contractor shall be responsible for the design, provision and maintenance of all temporary traffic signals and associated equipment unless otherwise given in the Contract.

## 505.7. Temporary lighting

Where required during the execution of the Works, the Contractor shall provide and maintain temporary lighting for the highways. Temporary lighting shall provide the same level of illumination as that of the existing street lighting it replaces. Temporary lighting shall be provided and approved prior to the removal of any existing street lighting.

## 505.8. Survey and reinstatement

1. Prior to commencing the Works, the Contractor shall carry out a condition survey of all roads and footways adjacent to the site. The survey record shall be available to the Contractor.

2. Unless stated otherwise, the Contractor shall reinstate all roads and footways affected by the Works to the extent, lines and levels that existed prior to the commencement of the Works and to standards that are at least equivalent to those that existed prior to the commencement of the Works.

3. Unless stated otherwise, the Contractor shall reinstate all surface water drainage systems (including but not restricted to gullies, channels, catchpits, piperuns, manholes and covers and the like) affected by the Works. The standard of reinstatement shall be at least equivalent to that existing prior to the Contract commencing.

## 505.9. Access within Works

1. The Contractor shall provide safe access in and about the site and underground workings and shall comply with the recommendations of BS 6164.

2. All shafts shall have a ladder access in addition to any mechanical means of access and egress.

3. The Contractor shall provide a safe designated pedestrian access in the tunnel and throughout the site area at all times. This shall have a firm, level, slip-resistant and continuous surface and shall be suitable for use in emergencies when lighting may be unavailable.

4. The Contractor shall segregate pedestrian and vehicular access routes.

5. The Contractor shall maintain a clear means of egress from each tunnel face at all times. Such means of egress through or past tunnelling machines, trains and similar obstructions shall meet the minimum dimensions in BS EN 16191.

**6.** The Contractor shall establish, maintain and operate a system whereby the presence of personnel underground is recorded, together with their location where appropriate.

## 506. Atmospheric testing

### 506.1. General

1. The parties to the Contract shall produce a schedule of gases and pollutants to be tested for, including both the frequency and the methods of testing. The recommendations of BS 6164 Section 4 shall be followed but this shall not restrict the range of testing to be carried out in any particular location.
2. No underground workings may be entered, except for rescue with appropriate respiratory protective equipment, where the oxygen content is below 19% by volume of the air present, or where explosive or noxious gases are present at concentrations in excess of safe limits.

### 506.2. Temperature

The Contractor shall avoid, if practicable, employing labour in areas where the wet bulb temperature is greater than 27°C. The recommendations of BS 6164 Clause 15.3 shall be applied. Where this requirement cannot be met, the Contractor shall devise and agree with the Engineer safe systems of work that take account of the exposure to heat or cold of the workforce.

## 507. Disposal of spoil and water

**507.1. Spoil waste programme**

The Contractor shall prepare a Site Waste Management Plan (SWMP) that sets out in detail how spoil and all waste is to be categorised, disposed of and monitored, and the programme for disposal and how legislation is to be complied with. This plan will address all waste matters at the site and have specific documented mechanisms for adopting a 'reduce, reuse, recycle' approach to waste minimisation for dealing with all wastes. The SWMP will be reviewed by the Engineer and accepted or approved as required by the Contract.

**507.2. Disposal of solid waste spoil**

The Contractor shall remove all excavated material, spoil, surplus materials and rubbish from whatever source on site and shall, except where otherwise specified in the Contract, make their own arrangements for its disposal and provide all the necessary facilities to achieve this. The Contractor shall also comply with any legal or local authority requirements applying to the handling and disposal of any contaminated spoil.

**507.3. Monitoring spoil removal**

1. The Contractor shall set up a system to control and monitor the transport of spoil from site to the tip site, in accordance with the current legislation. The system shall be agreed with the Engineer and will provide evidence that each load has been deposited at a licensed tip site.
2. The Contractor shall retain auditable records of waste removed from site. Waste Transfer Notices should be collated and submitted to the Engineer. Transfer and Consignment notes shall be kept in the site file.
3. The Contractor shall comply with all statutes and statutory instruments relating to spoil disposal.

**507.4. Liquid waste disposal**

1. The Contractor shall comply with the provisions of the Water Act 2003 where water is being pumped from the Works.
2. Before discharging any surplus water, the Contractor shall obtain the prior approval of the owner of the sewer or watercourse and of the regulatory authority.
3. The Contractor shall ensure that the condition of any discharged water complies with permitted limits. The parameters to be monitored include pH values, temperature and suspended solids, as a minimum.

# 508. Leakage

## 508.1. General

1. The acceptable degrees of leakage for tunnels, shafts and underground works may be specified in the Particular Specification (see Clause 508.1.3) according to the criteria in Table 15.

2. Specific criteria shall be defined in the Particular Specification for the junctions between the tunnels and other structures (cross-passages, shafts, adits and so on), taking into consideration the purpose of the ancillary structure and the type of connection.

3. For tunnels, shafts and underground works for which the Particular Specification does not require a specific measure of leakage or test, the Works shall be in accordance with the purpose of the structure, as defined in the column 'Typical use of the underground structure' of Table 15. Leakage that in the opinion of the Engineer is concentrated as defined in Clause 508.1.6 or significant or affects the use of the Works shall be sealed by the Contractor using approved methods and materials.

4. In the case of remedial works to correct or reduce water leakage, a specific assessment must be carried out to ensure that the works are not detrimental to the overall watertightness of the structure.

5. A damp patch or area may leave a slight film of moisture on the hand when touched briefly and for not more than 3 s, but no droplets of water or greater degrees of wetness are left on the hand.

6. Unless defined or agreed with the Engineer otherwise, concentrated leakage shall include

   (a) damp patches with an area greater than $4 \, m^2$
   (b) a total visible area of dampness greater than 10% of the visible tunnel face measured along a 10 m length of tunnel
   (c) local water flow in excess of 0.24 l/day channelled from an area of $1 \, m^2$ or a joint length of 1 m.

7. It should be noted that the degree of leakage defined here is not the same as the classifications provided in other standards such as BS 8102 Code of practice for protection of below ground structures against water from the ground (Table 2), and BS EN 1992-3 Design of concrete structures Part 3: Liquid retaining and containment structures (Table 7.105). Those standards define water tightness class or grade requirements while this document defines measurable criteria that may be used to meet those classes or grades.

## 508.2. Leakage criteria and classes of tunnel

1. The tunnel leakage criteria may be specified according to the degree of leakage as given in Table 15. The allowable daily leakage stated in Table 15 shall be taken as the average measured along the full length of each tunnel. For the purpose of this Clause the 'full length of each tunnel' shall be each length of tunnel, from one station or shaft or portal to the next. Local or concentrated leakage shall be as defined elsewhere or as stated in Clause 508.1.6 where specific measures are not defined.

2. The criteria of Table 15 apply only for seepage through the lining. They do not apply to planned drainage.

Table 15 Tunnel degree of leakage

| Degree of leakage | Dampness characteristics | Typical use of the underground structure | Definition | Allowable daily leakage (l/m²/day) |
|---|---|---|---|---|
| 1. | Absolutely dry | Storage rooms, residential, commercial and working areas with high sensitivity to water | No damp areas visible on the tunnellining | 0.01 |
| 2. | Substantially dry | Storage rooms and working areas in general | Occasional damp patches, which do not discolour blotting paper, detectable on the tunnel lining | 0.05 |
| | | Road tunnels with a risk of frost (unless mitigated by another measure) | | |
| | | Utility tunnels with high sensitivity to water | | |
| 3. | Capillary dampness | Rail tunnels without protection of the catenary systems, M&E & rails | Occasional damp patches on the tunnel lining, but no drops or trickle of water | 0.1 |
| | | Road tunnels with unprotected electrical equipment | | |
| | | Utility tunnels with unprotected electrical equipment | | |

Table 15 (Continued)

| Degree of leakage | Dampness characteristics | Typical use of the underground structure | Definition | Allowable daily leakage (l/m²/day) |
|---|---|---|---|---|
| 4. | Small amounts of dripping water | Rail tunnels provided that the catenary system, M&E systems & rails are protected from dripping water | Occasional drops or trickle of water | 0.2 |
| | | Road tunnels with protected electrical equipment, but no dripping water on the traffic lanes | | |
| | | Utility tunnels in general or with protected electrical equipment | | |
| 5. | Dripping or trickling water | Drainage and sewer tunnels | Occasional drops or trickle of water | 0.5 |
| | | | Local infiltration with low flow rate not impacting external environment or ground stability | |

## 508.3. Leakage tests

1. Where tunnels, shafts and underground works are required to perform to a specific measure of leakage, the work shall be acceptable only when successfully tested in accordance with the measures stated in the Contract. The time of testing shall be as stated in the Contract and shall take particular note of seasonal variations in water levels where appropriate. In the event that the test is unsuccessful the Contractor shall ascertain the areas where leakages occur and seal them using approved methods and materials and retest.

2. The testing shall include records of the following where considered relevant

   (*a*) testing method employed
   (*b*) tunnel temperature
   (*c*) external temperature and weather conditions
   (*d*) tunnel ventilation status and the approximate time at that status (for example, mechanically/not mechanically ventilated)
   (*e*) length of tunnel daily allowable leakage measured over, if not the full length.

The British Tunnelling Society
ISBN 978-0-7277-6643-4
https://doi.org/10.1680/st.66434.229
Published with permission by Emerald Publishing Limited under the CC BY-NC-ND 4.0 licence, https://creativecommons.org/licenses/by-nc-nd/4.0/

# Index

Printed in the USA
CPSIA information can be obtained
at www.ICGtesting.com
JSHW061917250124
56086JS00007B/184